# Student Study Guide with Selected Solutions



# ELECTRONICS

Fifth Edition

**Bill Deem**

By

**Barbara Bode Snyder**
*San Jose City College*

**Wallace McIntyre**
*Late, DeVry Institute of Technology, Chicago*

Prentice Hall

*Upper Saddle River, New Jersey* *Columbus, Ohio*

Editor: Stephen Helba
Production Editor: Christine M. Harrington
Cover Designer: Julia Zonneveld Van Hook
Production Manager: Pamela D. Bennett



Printed in the United States of America

10 9 8 7 6 5 4 3 2

ISBN: 0-13-577826-3

Prentice-Hall International (UK) Limited, *London*
Prentice-Hall of Australia Pty. Limited, *Sydney*
Prentice-Hall of Canada, Inc., *Toronto*
Prentice-Hall Hispanoamericana, S. A., *Mexico*
Prentice-Hall of India Private Limited, *New Delhi*
Prentice-Hall of Japan, Inc., *Tokyo*
Simon & Schuster Asia Pte. Ltd., *Singapore*
Editora Prentice-Hall do Brasil, Ltda., *Rio de Janeiro*

# PREFACE

Mathematics is an interesting and necessary adjunct to the electronics field. This guide/manual can be valuable as you prepare yourself for your study of electronics. This new and expanded edition contains a Student Study Guide and a Solutions Manual. After you study each chapter of the the text, you can strengthen your understanding of the material by reading the summary of the chapter as presented in this Study Guide.

Take the student quizzes at the end of each chapter. These questions are designed to test your skill and mastery of the chapter material. If you discover a weakness in your quiz results, go back and study the text. Retake the quizzes as needed and you will be successful in your mastery of the chapter material.

The Solutions Manual contains answers to all odd numbered end-of-chapter problems. Worked-out solutions to many problems, both odd and even, have been included to aid the student in his or her understanding of the problem-solving methods used. In many instances, more than one approach to a problem is possible and valid. The author has selected one and encourages students to try others. In problems where the solution can be obtained directly from the calculator with no other work necessary, only answers are given. For algorithms used in solving particular problem types, refer to the algorithms developed in the text.

**MATH** is a four-letter word. This four-letter word can have a special meaning.

**M** is you **MUST** know math for a good understanding of electronics.

**A** is be **ABLE** to work these problems and electronics will be easy for you.

**T** is to take the **TIME** to study math and your concept of electronic will become clearer.

**H** is **HAVE** a good attitude toward math and the four-letter word **MATH** will lead you to a successful career in electronics.

# CONTENTS

**Chapter 1** **THE DECIMAL NUMBER SYSTEM** **1**

Decimal Number System 1
Decimal Fractions 1
Rounding Whole Numbers 1
Rounding Non-Whole Numbers 2
Signed Numbers 2
Addition and Subtraction 2
Multiplication and Division 2
Mathematical Expressions and Terms 3
Practical Comments 3
Student Quiz 5

**Chapter 2** **POWERS OF TEN** **9**

Positive Exponents 9
Negative Exponents 9
Multiplication in Powers of Ten Form 9
Division in Powers of Ten Form 9
Combined Multiplication and Division in Powers of Ten 9
Scientific Notation 10
Problems with Complex Denominators 10
Reciprocals 10
Powers and Roots in Base Ten 10
Practical Comments 11
Student Quiz 13

**Chapter 3** **UNITS AND PREFIXES** **15**

Units 15
Prefixes 15
Applications 16
Systems of Measurement 16
Practical Comments 16
Student Quiz 17

**Chapter 4** **COMPUTER NUMBER SYSTEMS** **19**

Binary Number System 19
Octal Number System 19
Hexadecimal Number System 20
Binary to Octal to Hexadecimal Conversions 20
Addition 20
Subtraction 21
Practical Comments 21
Student Quiz 23

**Chapter 5** **ALGEBRAIC TERMS: ROOTS AND POWERS** **25**

Literal and Real Numbers 25
Algebraic Expressions and Terms 25
Numerical Coefficients 25
Exponents 25
Roots 26
Practical Comments 26
Student Quiz 27

Chapter 6 **FRACTIONS** **29**

Prime Numbers and Prime Factors 29
Reducing Fractions to Lowest Terms 29
Multiplication and Division of Fractions 29
Addition and Subtraction of Fractions 29
Improper Fractions and Mixed Numbers 30
Decimal Fractions 30
Practical Comments 30
Student Quiz 31

Chapter 7 **FRACTIONS AND LITERAL NUMBERS** **33**

Prime Numbers 33
Lowest Common Multiple 33
Multiplication and Division of Monomials 33
Multiplication and Division of Fractions 33
Addition and Subtraction of Fractions 33
Practical Comments 34
Student Quiz 35

Chapter 8 **LINEAR EQUATIONS** **37**

Identifying Equations 37
Linear Equations 37
Second-Degree Equations 37
Applications 37
Practical Comments 38
Student Quiz 39

Chapter 9 **FACTORING ALGEBRAIC EQUATIONS** **41**

Multiplication of Polynomials by Monomials 41
Multiplication of Binomials by Binomials 41
Division of Polynomials 41
Factoring Polynomials 42
Factors of Trinomials 42

Practical Comments 42
Student Quiz 43

Chapter 10 FRACTIONAL EQUATIONS 45
General Equations 45
Some Real Equations 45
Quadratic Equations 45
Practical Comments 46
Student Quiz 47

Chapter 11 DC CIRCUIT ANALYSIS: KIRCHHOFF'S LAWS 49
Kirchhoff's Current Law 49
Kirchhoff's Voltage Law 49
Polarity 49
Practical Comments 50
Student Quiz 51

Chapter 12 DC CIRCUIT ANALYSIS: OHM'S LAW 55
Circuit Resistance 55
Series Circuits 55
Parallel Circuits 56
Series-Parallel Circuits 56
Practical Comments 56
Student Quiz 57

Chapter 13 DC CIRCUIT ANALYSIS: CIRCUIT THEOREMS 61
Superposition Theorem 61
Thevenin's Theorem 61
Norton's Theorem 62
Practical Comments 62
Student Quiz 63

Chapter 14 **GRAPHING** 65

Graphing Linear Equations 65
Slope of a Line 65
Slope-Intercept Form 66
Interpreting Graphs 66
Plotting Curves 66
Practical Comments 67
Student Quiz 69

Chapter 15 **SIMULTANEOUS LINEAR EQUATIONS** 73

Linear Equations in Two Unknowns: Graphical Solution 73
Linear Equations in Two Unknowns: Algebraic Solution 73
Determinants 73
Applications 74
Practical Comments 74
Student Quiz 77

Chapter 16 **COMPLEX NUMBERS** 79

Imaginary Numbers 79
Complex numbers 79
Practical Comments 80
Student Quiz 81

Chapter 17 **THE RIGHT TRIANGLE** 83

Sides and Angles 83
Pythagorean Theorem 83
Trigonometric Functions 84
Trigonometric Tables 84
Trigonometric Equations 84
Practical Comments 84
Student Quiz 85

**Chapter 18** **TRIGONOMETRIC IDENTITIES** **87**

Law of Sines 87
Law of Cosines 87
Practical Comments 87
Student Quiz 89

**Chapter 19** **AC FUNDAMENTALS** **91**

Generating Angles 91
Generating Sine Waves 91
Frequency and Time 92
Angular Velocity 93
RMS and Peak Values of Voltage and Current 92
Phase Angle of Lead or Lag 92
Instantaneous Values and Voltage and Current 93
Practical Comments 93
Student Quiz 95

**Chapter 20** **AC CIRCUIT ANALYSIS: SERIES CIRCUITS** **97**

Series RC Circuits 97
Series RL Circuits 97
Polar to Rectangular Conversion 98
Series Resonance 98
Practical Comments 99
Student Quiz 101

**Chapter 21** **AC CIRCUIT ANALYSIS: PARALLEL CIRCUITS** **105**

RC Circuit Analysis 105
RL Circuit Analysis 105
Equivalent Circuits 105
AC Networks 106
Practical Comments 106
Student Quiz 107

Chapter 22 FILTERS 111

Low-Pass and High-Pass Filters 111
Thevenin's Theorem 112
Band-Pass Filters 112
Practical Comments 112
Student Quiz 113

Chapter 23 LOGARITHMS 117

Logarithms Defined 117
Common Logarithms 117
Multiplication of Numbers by Using Logarithms 117
Division of Numbers by Using Logarithms 118
Raising a Number to a Power by Using Logarithms 118
Natural Logarithms 118
Practical Comments 118
Student Quiz 119

Chapter 24 LOGARITHMIC EQUATIONS 121

Logarithmic and Exponential Forms 121
Logarithmic Equations: Common Logs 121
Logarithmic Equations: Natural Logs 121
Practical Comments 121
Student Quiz 123

Chapter 25 APPLICATIONS OF LOGARITHMS 125

Gain Measurement 125
Reference Levels 125
Frequency response 126
RC Circuits 126
Practical Comments 127
Student Quiz 129

**Chapter 26** **BOOLEAN ALGEBRA** **131**
Logic Variables 131
The AND Function 131
The OR Function 131
The NOT Function 132
Boolean Expressions 132
Booelan Postulates and Theorems 133
Practical Comments 134
Student Quiz 135

**Chapter 27** **KARNAUGH MAPS** **139**
Two-Variable Expressions 139
Three-Variable Expressions 140
Four-Variable Expressions 140
Practical Comments 140
Student Quiz 141

**Chapter 28** **INTRODUCTION TO STATISTICS** **143**
Population, Range, Mean, Median, Mode 143
Standard Deviation or Sigma 143
Normal Distribution Curve 144
Statistical Analysis 145
Practical Comments 146
Student Quiz 147

**ANSWERS TO STUDENT QUIZZES** **151**

**ANSWERS TO SELECTED PROBLEMS FROM TEXT** **155**

# THE DECIMAL NUMBER SYSTEM

CHAPTER 1

## DECIMAL NUMBER SYSTEM

- The word ***decimal*** means ten.
- The decimal system of numbers uses ten ***digits.*** A digit is a symbol for a number. We use these numbers in the decimal system: 0, 1, 2, 3, 4, 5, 6, 7, 8, and 9.
- Each position of a digit in a number has a ***place value.*** This is the number's value at that place.
- A digit that has the greatest effect upon the value of a number is called the ***most significant digit (MSD).***
- The ***least significant digit (LSD)*** is the digit with a place value that has the smallest effect upon the number's value.

## DECIMAL FRACTIONS

- A fraction has two numbers (1/2). The one above the line is called the ***numerator***, and the one below the line is the ***denominator.***
- A fraction whose denominator is 10 or a multiple of ten is known as a decimal fraction.
- To convert from a decimal fraction to a decimal number, determine the value of the denominator and place the LSD of the numerator in that position.
- Conversion from a decimal number to a fraction is the reverse process.

## ROUNDING WHOLE NUMBERS

- ***Rounding*** numbers to the nearest digit is a very common method used to simplify electronic calculations.
- A general rule is if the digit is 5 or larger, replace the digit with a 0 and add 1 to the next higher place value.

- To round down, a number 4 or less is replaced with a 0 and the next higher place value digit is lowered by 1.

## ROUNDING NON-WHOLE NUMBERS

- Use the same system as above to round decimal numbers. The number 0.745 would be rounded up to 0.75. Round 1.234 down to 1.23.

## SIGNED NUMBERS

- Numbers can have signs such as + or -. These numbers are called ***signed numbers***.

- Some common symbols are < and >. These mean ***less than*** and ***greater than***. For example, $1<2$ and $6.7>5.2$.

- If the value of the number is used without regard to the sign, then that value is called the ***absolute value***. The absolute value of -17 and 17 is 17.

## ADDITION AND SUBTRACTION

- The signs + and - indicate the sign of a number. They also indicate addition and subtraction.

- Adding two or more positive numbers involves just adding them and using a + sign on the result.

- Adding two or more negative numbers is a similar process to adding positive numbers. Just add the numbers and place a - sign before the result.

- Adding numbers with opposite signs is a slightly different process. Add all the positive numbers and then add all the negative numbers. Subtract the absolute values and place the sign of the largest number before the answer.

- The process of subtraction involves the first number called the ***minuend***, minus the second number, called the ***subtrahend***. The answer is called the difference.

- The rule in subtraction is to change the sign of the subtrahend and add as before.

## MULTIPLICATION AND DIVISION

- Symbols are used to denote multiplication and division. The symbol for multiplication is X or • (i.e., 3 X 4 and 3 • 4 are the same). Division symbols are ÷ or /(i.e., 4/5 and 4÷5).

- Multiplying or dividing signed numbers requires a method of determining the sign of the result.
- Positive signed numbers multiplied or divided will always result in a positive answer.
- Multiplying or dividing an even number of negative numbers will result in a positive answer.
- If the number of negative numbers multiplied or divided is odd, then a negative answer is correct.

## MATHEMATICAL EXPRESSIONS AND TERMS

- A mathematical ***term*** is a number preceded by a + or - sign.
- An ***expression*** is a group of two or more terms.
- There is a definite order required to solve a mathematical expression. Perform the multiplications and divisions first and then perform the additions and subtractions.
- Mathematical expressions are often grouped in together in parenthesis ( ), brackets [ ], or braces { }.
- A mathematical expression using these groupings must be solved by performing the calculations (x, ÷, +, or -) in each set of parentheses, brackets, or braces first. Then combine the results from within the parentheses, brackets, or braces, keeping in mind the proper order of performing multiplications and divisions followed by additions and subtractions.

# PRACTICAL COMMENTS

- You are on your way to making a career in electronics. The electronics field requires a method of thinking and approaching a problem with tenacity and learning. This problem solving cannot be achieved in a short period. Time and practice are necessary. Your study of math will help you to gain the learning necessary to operate with efficiency in your work with electronic circuits.

- The use of math in analyzing electronic circuits will show you the way that all circuits behave under any conditions. As an example, you have just installed a new battery in your car. The engine still does not turn over fast enough to start the car. Remembering Ohm's law (which taught you the relationship between voltage, current and resistance), you speculate that if the voltage is high (because of the new battery) and the starter is almost new, then high resistance must be causing the motor to slow. Bingo, you forgot to clean the battery terminals (the source of high resistance). You clean them, and away you go. This type of logical thinking is helped by your study of math.

## STUDENT QUIZ

Name ______________________________________________

1. A mathematical expression is made up of one or more terms.
   a. true
   b. false
2. The correct order of performing operations in an expression is +, -, x, and ÷.
   a. true
   b. false
3. MSD means the most significant digit.
   a. true
   b. false
4. In the place value table for the decimal number system the third place has a value of hundreds.
   a. true
   b. false
5. Zero is not a digit in the decimal system.
   a. true
   b. false
6. Convert 746 ten-thousandths to a decimal fraction.
   a. 74.6/10,000
   b. 746/1,000
   c. 10,000/746
   d. 746/10,000
7. Given the number 567.89, the MSD is ________ and the LSD is ________.
   a. 9, 5
   b. 5, 8
   c. 5, 9
   d. 6, 7
8. Convert 438 thousandths to a decimal number.
   a. 0.438
   b. 0.0438
   c. 0.00438
   d. 0.000438
9. Convert 31/10,000 to a decimal number.
   a. 0.31
   b. 0.031
   c. 0.0031
   d. 0.00031
10. Round the number 12,965 to the nearest hundred.
    a. 12,966
    b. 12,900
    c. 12,010
    d. 13,000

11. Round the number 7.654 to two decimal places.
  a. 7.7
  b. 7.65
  c. 7.64
  d. 7.66
12. The statement $8.9<9.1$ is _________, and the statement $0.73>0.74$ is _________.
  a. true, true
  b. true, false
  c. false, false
  d. false, true
13. The absolute value of 17.5 is _________, and the absolute value of -0.56 is _________.
  a. 17.5, -0.56
  b. -17.5, -0.56
  c. 17.5, 0.56
  d. none of these
14. Perform the following. $30 + (-40) - (-45)$
  a. 35
  b. -35
  c. -55
  d. 115
15. Given this expression, $21 - (-35) - 67$. Find the absolute value of the answer.
  a. -11
  b. 11
  c. 81
  d. -81
16. Perform the following. $-67-(-34)+(-8)-(-12)-29$
  a. -29
  b. -58
  c. -55
  d. -97
17. Perform the following. $(3-12)+15\div5-7\text{x}2$
  a. -2
  b. -5
  c. -20
  d. -32
18. Perform the following. $\{12 + (-3\text{x}-12)\div(2\text{x}6)\}$
  a. 15
  b. -15
  c. 9
  d. -9

19. Perform the following. [64-(4x8)][84/7+12]
    a. 396
    b. 768
    c. -308
    d. -396

20. Perform the following. {18-(3x7)+12}÷{4x6÷8}
    a. 21
    b. 3
    c. 0
    d. 9

# POWERS OF TEN

CHAPTER 2

## POSITIVE EXPONENTS

- Large or small numbers can be handled easily by using a system of notation called ***powers of ten.***
- The powers of ten refer to the ***exponents*** of ten. This exponent is the number of times ten is multiplied by itself. For example $10 \times 10 \times 10 \times 10 = 10^4$. The four is the exponent of ten.
- To convert from a powers of ten to a number, simply write 1 and follow it by the number of 0's in the exponent.

## NEGATIVE EXPONENTS

- Numbers less than 0 are dealt with by using negative exponents.
- Negative exponents determine the number of places the decimal point is moved from the present decimal point towards the left to the new decimal point. An example of this is $10^{-3} = 0.001$.

## MULTIPLICATION IN POWERS OF TEN FORM

- Multiplication of terms with powers of 10 is simply a matter of adding the powers. $10^5 \times 10^3 = 10^8$

## DIVISION IN POWERS OF TEN FORM

- Division of numbers containing powers of ten involves subtraction of the exponents. $10^3/10^2 = 10^1$

## COMBINED MULTIPLICATION AND DIVISION IN POWERS OF TEN

- This process is one of adding or subtracting the powers of ten as required.
- The powers of ten are added or subtracted using the same rules as adding or subtracting signed numbers.

- To move a power of ten from a denominator to a numerator, simply change the sign of the exponent. $1/10^3 = 10^3$. The opposite change is also true.

## SCIENTIFIC NOTATION

- A very convenient method of expressing the numbers we use in electronics is to use ***scientific notation***.
- Scientific notation is the placing of a decimal point after the MSD and adding a power of ten. The exponent of the power of ten will be the number of places the decimal point is moved.
- Count the number of decimal places the decimal point moves left. This is the positive value of the exponent.
- Move the decimal point to the right and the exponent is negative.
- Use your scientific calculator to do problems involving scientific notation.
- Your calculator can handle any calculations with powers of ten.

## PROBLEMS WITH COMPLEX DENOMINATORS

- If your problem has a complex denominator, that is with numbers added or subtracted, remember that the denominator operation must be done first.
- If you have a calculator that has a parenthesis feature, then this is a simple problem.

## RECIPROCALS

- A ***reciprocal*** of a number is one divided by that number. As an example, 1/2 is the reciprocal of 2.
- Most calculators have a separate key to give you the reciprocal of any number in its display.

## POWERS AND ROOTS IN BASE TEN

- To raise a power of ten to a power simply multiply the exponents. $(10^2)^{-2} = 10^{-4}$
- To find the root of a number, you can raise that number to the reciprocal of the exponent. For example, $(10^6)^{1/2}$ will give you the square root of $10^6 = 10^3$.

# PRACTICAL COMMENTS

- The numbers used in the electronics field are often very large or very small. Using the powers of ten is the most convenient method of expressing these large or small numbers. For example, electrical current is often measured in milliamps, (mA). This is the same as $10^{-3}$ or 0.001 amperes. The other extreme is a common unit of resistance called the ohm ($\Omega$). This value can be several thousand ohms or even several million ohms. 1,000,000 ohms is the same as $10^6\ \Omega$. This is abbreviated as 1 Megohm or even shorter, 1 M$\Omega$. Another electrical quantity is capacitance measured in micro farads ($\mu$F). 1 $\mu$F = 1/1,000,000 Farad or $10^{-6}$ F. Powers of ten will be used constantly in your chosen field of electronics.

- Electronic engineers and technicians have a very valuable tool constantly at hand, the calculator. Most electronic happenings in a circuit can be explained by mathematical processes. The hand held calculator excels at solving these equations easily and accurately. By the way, the calculator is very handy for checking the numbers on your pay checks.

## STUDENT QUIZ

Name ______________________________

1. Powers of ten are a convenient way of expressing only large numbers.
   a. true
   b. false
2. The calculator is very handy in electronic math.
   a. true
   b. false
3. Powers of ten can be converted to a decimal number by moving the decimal point either to the right or left depending on the sign of the exponent.
   a. true
   b. false
4. The reciprocal of 10 is 0.01.
   a. true
   b. false
5. Scientific notation is used in electronics to simplify and shorten mathematical expressions.
   a. true
   b. false
6. Convert $10^4$ to a decimal number.
   a. 1
   b. 10
   c. 100
   d. 1000
   e. 10000
7. Convert $10^{-6}$ to a decimal number.
   a. 0.0001
   b. 0.00001
   c. 0.000001
   d. 0.0000001
8. Convert $10^{-3}$ to a decimal fraction.
   a. 1/10
   b. 1/100
   c. 1/1000
   d. 1/10000
9. Convert 0.001 to a power of ten form.
   a. $10^{-2}$
   b. $10^{-3}$
   c. $10^{-4}$
   d. $10^{-5}$

10. Perform the following operation. $10^{-5} \times 10^{3}$
   a. $10^{-2}$
   b. $10^{2}$
   c. $10^{-8}$
   d. $10^{8}$
11. Divide the following. $10^{0}/10^{3}$
   a. 10
   b. 1
   c. $10^{3}$
   d. $10^{-3}$

12. Perform the following. $(10^{3} \times 10^{-1} \times 10^{5}) \div (10^{3} \times 10^{5})$
   a. $10^{-0}$
   b. $10^{-1}$
   c. $10^{-3}$
   d. $10^{-5}$
13. Given this number, 5855. Round this to two significant places and express your answer in scientific notation.
   a. $5.855 \times 10^{3}$
   b. $5.86 \times 10^{3}$
   c. $5.9 \times 10^{3}$
   d. $5.9 \times 10^{-3}$
14. Solve for the value of this expression, 1/47000. Express your answer in scientific notation and round to two decimal places.
   a. $2.12 \times 10^{-5}$
   b. $2.13 \times 10^{-5}$
   c. $2.127 \times 10^{-4}$
   d. $2.1276 \times 10^{-7}$
15. Solve the following. $1/[(1/(4.7 \times 10^{3}) + 1/(2.2 \times 10^{3}]$
   a. $1.499 \times 10^{4}$
   b. $6.673 \times 10^{-5}$
   c. $6.9 \times 10^{4}$
   d. $1.449 \times 10^{-5}$

# UNITS AND PREFIXES

CHAPTER 3

## TS

Converting from one power of ten to another is not difficult. To change to a power of ten which is more positive, compensate by moving the decimal point in the numerical part to the left to make the numerical part smaller.

$5.678 \times 10^2 = .5678 \times 10^3 = .05678 \times 10^4 = .0005678 \times 10^6$

To change to a power of ten which is less positive (more negative), compensate by moving the decimal point in the numerical part to the right to make the numerical part larger.

$.0005678 \times 10^6 = .5678 \times 10^3 = 567.8 \times 10^0 = 567800 \times 10^{-3}$

A number expressed with a single digit to the left of the decimal point and times a power of ten is expressed in ***scientific notation***, such as $5.23 \times 10_4$.

***Engineering notation*** is a number expressed in two parts, a numerical part and a power of ten where the power of ten is a multiple of 3 or -3. The decimal point can be anywhere in the numerical part, such as $45.67 \times 10^3$. The power of ten can be replaced by a prefix and still be called ***engineering notation***.

Converting from a number to ***scientific notation*** is easy. Just move the decimal point to the right of the most significant digit and add a power of ten. If you moved it left (in the negative direction) 4 places, then the exponent on the power of ten must be +4. If you moved it right (in the positive direction) 2 places, then the exponent on the power of ten must be -2.

$47000\ \Omega = 4.7000 \times 10^4\ \Omega$ (if you don't round it).

$.01234\ A = 1.234 \times 10^{-2}\ A$

## FIXES

***Engineering notation*** used in electronics has abbreviations for powers of ten which are multiples of 3 and -3. The most important of these are $10^6$ = Mega, $10^3$ = kilo, $10^{-3}$ = milli,

$10^{-6} = \mu$ = micro [note: $\mu$ is the lower case Greek letter mu],

$10^{-9}$ = nano, $10^{-12}$ = pico, $10^{-15}$ = femto.

Remember that any number to zero power $= 1 = 10^0 = 2^0 = 8^0$

The basic electrical quantities also have their abbreviations. A resistor used in series has resistance is expressed in Ohms ($\Omega$) and used in parallel has conductance measured in Siemens (S). Current is expressed in amperes (A), voltage in volts (V), and power in watts (W).

## APPLICATIONS

- Calculations involving electronic quantities are commonly expressed in engineering units ar abbreviated. Be sure to keep as many significant digits as were in the original data. Fo example,
  2.2 mV / 4.7 kΩ = 5.0 x $10^{-7}$ A = .50 x $10^{-6}$ A = .50 $\mu$ A

## SYSTEMS OF MEASUREMENT

- Electronics uses units to measure many quantities that are very large or very small.

- These units of measure are based on a system called the ***Internation System of units (SI***

- Unit names are abbreviated, as capacitance (C), current (I), conductance (G), resistance (R electromotive force (E or V), power (P).

- Prefixes related to powers of ten which are multiples of 3 or -3 are used with the units express quantities of measure such as
  34.5 x $10^{-3}$ V = 34.5 mV = .00345 V

# PRACTICAL COMMENTS

- All people in electronics use abbreviations. Students new to electronics must becor familiar with these new terms. Changing from amperes to milliamperes (mA), for exampl is an everyday occurrence. Some of you will arrive at answers of 4.56 x $10^4$ Ω on yo calculator. While this is correct, it is more convenient to be able to express the answer kΩ. Since k = $10^3$ , to change from $10^4$ to $10^3$ is to make the power of ten one less positiv so to compensate move the decimal point one place in the direction to make the numeric part more positive.
  4.56 x $10^4$ Ω = 45.6 x $10^3$ Ω = 45.6 kΩ = 45600 x $10^0$ Ω = 45600 Ω

- Practice these conversions well and your study of electronics will be much smoother.

- Series circuits use ohms (Ω) values. Parallel circuits use Siemens values. Usually Ω a Siemens are reciprocals ( 1/X or $X^{-1}$ ) of each other. Do NOT say inverse for reciproc
  41 Ω = ( 1/41) S = .02439 S = 24.39 mS
  47 kΩ = 47000 Ω = ( 1/47000 ) S = 0.00002128 S = 21.28 $\mu$S

## STUDENT QUIZ

Student Name ______________________________________

1. Engineering notation is the expression of a number times a power of ten in multiples of 3.
   a. true
   b. false
2. Scientific notation is expressing any number times a power of ten.
   a. true
   b. false
3. Some typical electronic units are volts, ohms, and amperes.
   a. true
   b. false
4. The prefix micro = $10^{-3}$.
   a. true
   b. false
5. The correct power of ten to express Mega is $10^{6}$.
   a. true
   b. false
6. Express the number 0.0067 A as ________ mA and as _______ $\mu$A.
   a. 6.7, 0.067
   b. 67, 670
   c. 6.7, 6700
   d. 0.67, 0.67
7. Express the number 1.23 V as _______ mV and as ______ kV.
   a. 1.23, 0.123
   b. 1230, 0.00123
   c. 12.3, 0.00123
   d. 1230, 0.123
8. Change 14,976 Hz to kHz.
   a. $1.4976 \times 10^{5}$ kHz
   b. 14.976 kHz
   c. 147.96 kHz
   d. 1479.6 kHz
9. Express the number $4.7 \times 10^{-5}$ to $\mu$F and then in pF.
   a. 4.7 $\mu$F, 4700 pF
   b. 0.047 $\mu$F, 47000 pF
   c. 470 $\mu$F, 470000 pF
   d. 47 $\mu$F, 47000000 pF
10. Express 2.2 k$\Omega$ as $\Omega$.
    a. 22 $\Omega$
    b. 220 $\Omega$
    c. 2200 $\Omega$
    d. 22000 $\Omega$

11. Express 0.005 μF as pF.
    a. 5 pF
    b. 50 pF
    c. 500 pF
    d. 5000 pF
12. Change 14 MHz to Hz.
    a. 14,000,000 Hz
    b. 1,400,000 Hz
    c. 140,000 Hz
    d. 14,000 Hz
13. Perform the indicated calculation. 15 V/4.7 kΩ =
    a. 319 μA
    b. 313 mA
    c. 0.0313 A
    d. 3.19 mA
14. Perform the following. 1/(6.28 x 12 kHz x 0.47 μF)
    a. 28.2 Ω
    b. 28.2 kΩ
    c. 0.035 Ω
    d. 0.035 kΩ
15. Perform the following. $1/6.28(0.47 \text{ mH} \times 0.47\ \mu\text{F})^{1/2}$
    a. 03.3 μHz
    b. 10.7 kHz
    c. 494 kHz
    d. 494 Hz

# COMPUTER NUMBER SYSTEMS

CHAPTER 4

## BINARY NUMBER SYSTEM

- The ***binary number system*** has only two digits, 0 and 1.
- In the decimal system the LSD has a value of $10^0$. The next least significant digit is $10^1$, and so on. The binary system follows the same system. The base is two instead of the decimal's 10.
- The place values for the binary places are $2^0 = 1$, $2^1 = 2$, $2^2 = 4$, and $2^3 = 8$ and so on.
- Binary digits are called ***bits***.
- In the binary system we have the ***least significant bit (LSB)*** and the ***most significant bit (MSB)***.
- To convert from a binary number to a decimal number, add the place values of each place where the binary digit is 1. As an example, $1101_2 = 8 + 4 + 0 + 1 = 13_{10}$.
- Conversion from decimal to binary uses a continual divide-by-2 method. Divide the decimal number by 2. The remainder if any will be a 1. The remainder of 0 or 1 is the LSB of the binary number. Divide the whole number, answer to your first division above by two. If the remainder is 1, that is the next bit. If there is no remainder, then 0 will be the bit. Continue dividing the answers to previous divisions until no further divisions can be made.

## OCTAL NUMBER SYSTEM

- Another number system that finds some use in digital circuits is the ***octal number system***.
- The base of this system is 8.
- The octal digits are 0, 1, 2, 3, 4, 5, 6, and 7.
- The octal system has, as we have seen before, an LSD of $8^0$ which equals 1. The next digit's place value is $8^1 = 8$ and so on.

- Reading an octal number, such as, $160_8$ is one-six-zero not one hundred and sixty.

- Converting from octal to decimal involves the corresponding place value table we used in binary to decimal conversion.

- Multiply the octal digit's place value by the octal number and add the results. For example,
  $734_8 = (7 \times 8^2) + (3 \times 8^1) + (4 \times 8^0) = 448 + 24 + 4 = 476^{10.}$

- Decimal to octal conversion can be done by dividing successively by 8 with the remainder being the octal number. Remember the first division will produce the octal LSD.

## HEXADECIMAL NUMBER SYSTEM

- A number system with a base of 16 is the ***hexadecimal number system***.

- This system is often abbreviated hex.

- The hex system has 16 digits 0, 1-9, and A, B, C, D, E, and F.

- Hex to decimal and decimal to hex conversions are similar to our other conversions.

## BINARY TO OCTAL TO HEXADECIMAL CONVERSIONS

- Binary to octal conversions are made by starting at the LSB of the binary number and dividing the bits into groups of 3. Then write the octal digits for each binary group. If the binary number to be converted is 10110111011, the rightmost 3 bits = 3, the next group of 3 bits = 6, and the left most 2 bits = 2. The result, therefore is $2673_8$.

- Conversions from binary to hexadecimal is a similar process. Divide the binary number into groups of four bits. Be sure to start at the LSB. Substitute the hex digit for each group.

- Conversions between any two of our systems can be accomplished by use of a calculator designed for this operation. Calculators have a limit on the size of numbers they can store. This is especially true of the binary system since large numbers expressed in binary use the most digits.

## ADDITION

- Adding decimal numbers involves an operation known as the carry. That is, when the sum of a column is greater than 9 a carry is placed in the next column.

- Octal numbers can also be added with a carry value of 8. This carry value must be subtracted from the column sum. This subtracted answer is the octal place value. Of course, you then carry a 1 to the next column.

- Hexadecimal numbers are added by using a carry value of 16. This is similar to binary and octal addition.

**SUBTRACTION**

- Decimal numbers are subtracted by using the borrow. We are very familiar with this system.

- Subtracting octal numbers involves the same borrow procedure. The value of the borrow is 8.

- Hexadecimal numbers use a borrow of 16.

- Binary numbers can also be subtracted by using the borrow process.

- The use of the ***one's complement*** method of subtracting binary numbers is generally easier than using the borrow method.

- Finding a one's complement of a binary number is simply changing all the 1's to 0's and vice versa.

- A one's complement is used to subtract as follows. Change the subtrahend to a one's complement. Add the one's complement to the minuend. Perform an end around carry and add again. An end-around carry is adding the final MSB carry.

- The use of a ***two's complement*** to subtract binary numbers is very popular in computers.

- A two's complement is a one's complement plus a 1. For example, the one's complement of $101_2$ is 010. By adding a one you get the two's complement: $010 + 1 = 011 = 011$ which is the two's complement of $101_2$

- To subtract using the two's complement, add the two's complement of the subtrahend to the minuend and ignore the carry.

## PRACTICAL COMMENTS

- A very simple way of converting hex or octal to decimal is to convert each number to binary. Then a conversion to decimal is easy. Unless you regularly make a lot of conversions you will remember this system longer than any other method. Fewer mistakes are made by students using this shortcut.

- Use the calculator every chance you get to do these types of calculations. You need to know the manuald methods but later when you go into the working world, the calculator will be your quick solution.

## STUDENT QUIZ

Name ______________________________

1. The binary number system uses two digits, 0 and 1.
   a. true
   b. false
2. The octal system uses the digits 1, 2, 3, 4, 5, 6, 7, and 8.
   a. true
   b. false
3. Hexadecimal numbers are commonly used in computer systems because they have fewer digits.
   a. true
   b. false
4. The two's complement method is a way to add binary numbers.
   a. true
   b. false
5. The digits used in the hexadecimal number system include those used in the decimal system plus A, B, C, D, and E.
   a. true
   b. false
6. Convert this number, $10010110_2$ to decimal.
   a. $151_{10}$
   b. $51_{10}$
   c. $150_{10}$
   d. $351_{10}$
7. Convert the number $245_{10}$ to binary.
   a. $101100011_2$
   b. $110011010_2$
   c. $110000111_2$
   d. $11110101_2$
8. Convert $245_8$ to base 10.
   a. $165_{10}$
   b. $821_{10}$
   c. $220_{10}$
   d. $123_{10}$
9. The hexadecimal number, $2C80_{16}$ expressed as a binary number is
   a. $0010\ 1110\ 1000\ 0000_2$
   b. $0001\ 1100\ 1000\ 0000_2$
   c. $0010\ 1100\ 1000\ 0000_2$
   d. $0000\ 1000\ 1100\ 0010_2$
10. Convert the following binary number, $110001110010_2$ to an octal number.
    a. $6161_8$
    b. $6162_8$
    c. $2616_8$
    d. $6261_8$

11. Add the following $1100110_2 + 101010_2$.
   a. $10010000_2$
   b. $10010010_2$
   c. $10001110_2$
   d. $10011101_2$

12. Add the following. $ABCD_{16} + 10C5_{16}$
   a. $ABD3_{16}$
   b. $CDE7_{16}$
   c. $ABD2_{16}$
   d. $BCD2_{16}$

13. Find the two's complement of $10010010_2$.
   a. 01101101
   b. 01101110
   c. 10011001
   d. 01101001

14. Perform the following. $1100110_2 - 10110_2$
   a. $1010000_2$
   b. $1111100_2$
   c. $1011100_2$
   d. $1010_2$

15. Perform the following. $8B4C_{16} - 10FC_{16}$
   a. $F4A0_{16}$
   b. $9C48_{16}$
   c. $3418_{16}$
   d. $7A50_{16}$

# ALGEBRAIC TERMS: ROOTS AND POWERS

CHAPTER 5

## LITERAL AND REAL NUMBERS

- A ***real number*** is the kind of number we have always dealt with, such as 17, 5.7, or -2.4.

- A real number can have only one value.

- Letters are often used to represent a variable value. These are called ***literal numbers***.

## ALGEBRAIC EXPRESSIONS AND TERMS

- An algebraic expression can contain one or several combinations of both real and literal numbers.

- An algebraic expression with one term is called a ***monomial***.

- An algebraic expression containing more than one term is called a ***polynomial***.

## NUMERICAL COEFFICIENTS

- If a real number precedes an algebraic term then that number is known as the ***numerical coefficient***. As an example, in the term 3ab, 3 is the numerical coefficient.

- If no number is present, then the coefficient is assumed to be 1, as in the term abc.

## EXPONENTS

- An exponent is, as you will recall, a number that tells us how many times that number is multiplied by itself. Such as, $5^3 = 5 \times 5 \times 5 = 125$

- When dealing with a literal number, the value cannot be found until the value of the literal number is known. The expression $a^2 = a \times a$ cannot have a value until you know the value of a.

- The calculator is most useful for determining the values of numbers with exponents.

## ROOTS

- The ***square root*** of a number is one of its two equal factors. 7 x 7 = 49 is an example, where 7 is the square root of 49.

- The radical sign (√), denotes the square root.

- A square root is also expressed as an exponent, such as $\sqrt{2} = 2^{1/2}$

- Use a calculator for most of these types of calculations.

## PRACTICAL COMMENTS

- The use of literal numbers is most prevalent in the field of electronics. Formulas are expressed in literal terms so that they can apply to a variety of circuit conditions. For example, Ohm's Law is expressed as $I = V/R$. If you know a value for I and R, then you can calculate for the value of V. Another example is the combination of real and literal numbers. The expression for determining the value of inductive reactance in an AC circuit is $X_L = 2\pi fL$. In this case the 2 and the $\pi$ are real numbers and the f (frequency) and the L (inductance) are literal values.

## STUDENT QUIZ

Name ______________________________

1. In the expression 6ab, the 6 is a literal number.
   a. true
   b. false
2. The square root of a number is that number multiplied by itself.
   a. true
   b. false
3. The expression $a^3 = 8$ is correct if $a = 2$ .
   a. true
   b. false
4. The square root of a number can be found by raising the number to the 1/2 power, such as $2^{1/2}$.
   a. true
   b. false
5. In the expression $x^4$, the 4 is known as an exponent.
   a. true
   b. false
6. Calculate the following and round to three significant figures. $1.2^4$
   a. 1.44
   b. 1.73
   c. 2.07
   d. 2.49
7. Solve the following if $a = 3$. $a^5$
   a. 27
   b. 81
   c. 243
   d. 729
8. If $a = 1$, $b = 2$, and $c = 3$, find the value of the expression $abc^2$.
   a. 9
   b. 18
   c. 27
   d. 36
9. If $a = 1$, $b = 2$, and $c = 3$, find the value of the expression $4(abc)^3$.
   a. 108
   b. 216
   c. 432
   d. 864
10. If $a = 1$, $b = 2$, and $c = 3$, find the value of the expression $3a^2 - b^3 + 4c^2$.
    a. 31
    b. 19
    c. 27
    d. 42

11. Solve and round the following to three significant figures. $38^{1/2}$.
    a. 6.164
    b. 6.16
    c. 2.48
    d. 1440
12. Solve and round the following to three significant figures. $289^{1/2}$.
    a. 17.6
    b. 22
    c. 49
    d. 17
13. Solve and round the following to three significant figures. $99^{1/2}$
    a. 9.95
    b. 9800
    c. 0.995
    d. 0.00995
14. If $a = 1$, $b = 2$, and $c = 3$, find the value of the expression $(a^2 + 2b^2 + 3c^2)^{1/2}$
    a. 19
    b. 24
    c. 36
    d. 6
15. If $a = 1$, $b = 2$, and $c = 3$, find the value of the expression $(2a^2 + 2a^2b^2 + 2a^2b^2c^2)^{1/2}$ Round to three significant numbers.
    a. 82
    b. 9.06
    c. 9
    d. 8.49

# FRACTIONS

CHAPTER 6

## PRIME NUMBERS AND PRIME FACTORS

- A ***prime number*** is a number that has no whole-number factors except one and itself. An example of a prime number is 7.

- A non-prime number can be divided by prime numbers. These prime numbers are called ***prime factors***. An example is that 2 and 5 are prime factors of 10.

## REDUCING FRACTIONS TO LOWEST TERMS

- Often when adding or subtracting fractions it is convenient to find the prime numbers of two or more numbers. Finding these results in determining the ***lowest common multiple (LCM)***.

- To find the LCM of two or more numbers, find the prime numbers of each number. The LCM is then the various prime numbers found multiplied by the number of times each prime is used. For example, to find the LCM of 10 and 12, determine the prime factors: $10 = 2 \times 5$ and $12 = 2 \times 2 \times 3$. The factors 5 and 3 occur once and 2 occurs twice. From this, multiply as follows: LCM $= 5 \times 3 \times 2 \times 2 = 60$.

## MULTIPLICATION AND DIVISION OF FRACTIONS

- Fractions must be reduced to the lowest terms. This occurs when there are no like factors in the numerator or denominator.

- To multiply fractions simply multiply the numerators together. Then multiply the denominators. Reduce the result to lowest terms.

- Division of fractions involves inverting the denominator and multiplying as above.

## ADDITION AND SUBTRACTION OF FRACTIONS

- Two or more fractions requiring addition must have a common denominator(CD) before they can be added.

- If two or more fractions have a common denominator, the numerators can be added together.

- Subtraction of fractions is similar to addition as long as a common denominator is found.

## IMPROPER FRACTIONS AND MIXED NUMBERS

- An ***improper fraction*** is a fraction where the numerator is larger than the denominator.

- ***Mixed numbers*** are numbers that have a whole number part and a fraction part, such as 4 1/3.

- To change an improper fraction to a mixed number, divide the numerator by the denominator to get the whole number. The remainder is the numerator of the fraction.

- Division or multiplication of improper fractions follows the same rules as ordinary fractions.

- Adding and subtracting mixed numbers involves changing all the numbers to improper fractions. Then it is easy to find the common denominator and proceed as above.

## DECIMAL FRACTIONS

- The most common solution to expressing fractions as decimal fractions is to use your calculator. To convert 11/15 to a decimal fraction, divide 11 by 15 and round the answer to the desired accuracy.

# PRACTICAL COMMENTS

- Many electronic formulas involve fractions. For example we refer again to Ohm's Law. $I = V/R$. The term V/R is a fraction which shows the relationship of voltage to resistance in any circuit. If $V = 20$ V and $R = 10\ \Omega$, then the fraction $20\text{ V}/10\Omega = 20/10 = 2$ Amperes. This fraction also tells us that any combination of V/R that equals the whole number 2 will always produce 2 amperes of current in the circuit.

## STUDENT QUIZ

Name ______________________________

1. An example of a prime number is 15.
   a. true
   b. false
2. The prime factors of a non-prime number are those prime numbers that can be divided into the non-prime number.
   a. true
   b. false
3. To multiply fractions simply add the numerators and multiply the denominators together.
   a. true
   b. false
4. To divide fractions, invert the denominator and multiply.
   a. true
   b. false
5. A mixed number has a whole number and a fraction.
   a. true
   b. false
6. Reduce the following fraction to lowest terms. 26/208
   a. 1/16
   b. 2/64
   c. 1/8
   d. 2/16
7. Multiply the following and reduce to the lowest terms. 2/3 x 2/5 x 2/7
   a. 4/15
   b. 14/15
   c. 4/35
   d. 8/105
8. Divide the following and reduce to the lowest terms. 9/10 ÷ 4/5
   a. 1 1/8
   b. 7/8
   c. 8/9
   d. 18/25
9. Change the following to a mixed number. 160/9
   a. 17.778
   b. 17 7/9
   c. 8 8/19
   d. 160/18
10. Change the following to an improper fraction. Reduce to the lowest terms. 6 9/16
    a. 8/27
    b. 27/8
    c. 105/16
    d. 3 1/8

11. Perform the following and express your answer as a mixed number. 7 2/5 + 3 3/8
    a. 5 15/20
    b. 1 9/16
    c. 10 8/9
    d. 10 31/40
12. Perform the following and express your answer in decimal numbers. Round to three decimal places. 7 3/5 x 9 2/3
    a. 73.467
    b. 17.267
    c. 72.467
    d. 27.469
13. Perform the following and reduce to the lowest terms. 7/30 - 9/10
    a. 21/100
    b. 1 2/15
    c. 7/27
    d. -2/3
14. Perform the following and change the answer to a mixed number. 4 1/8 ÷ 2 3/16
    a. 1 31/35
    b. 66/35
    c. 9 2/128
    d. 1 15/16
15. Perform the following and express the answer as a mixed number. 2 1/8 x 4 3/5
    a. 6 29/40
    b. 2 19/40
    c. 9 31/40
    d. 85/184

# FRACTIONS AND LITERAL NUMBERS

CHAPTER 7

## PRIME NUMBERS

- A literal number is a prime number when it is raised to the first power. A literal number raised to another power is not prime. For example, y is prime while $y^2$ is not prime, since $y^2 = y \cdot y$

## LOWEST COMMON MULTIPLE

- To find the LCM of two literal numbers, factor each term, find the maximum number of times a prime factor appears in any one term, and multiply the primes found. This process is similiar to finding the LCM or LCD of two numbers.

## MULTIPLICATION AND DIVISION OF MONOMIALS

- Multiplication of monomials containing literal numbers involves multiplying the real and then the literal numbers.

- Division of monomials containing literal numbers involves the division of the real numbers followed by division of the literals. For example $6a^2/3a = 2a$.

## MULTIPLICATION AND DIVISION OF FRACTIONS

- To multiply fractions containing literal numbers just multiply the numerators and then the denominators. Cancel like factors and reduce to the lowest terms.

- To divide fractions containing literal numbers, invert the denominator of one of the fractions and multiply the fractions.

## ADDITION AND SUBTRACTION OF FRACTIONS

- If the denominators of the two fractions to be added or subtracted are equal, just add or subtract the numerator.

- You must find a common denominator before fractions can be added or subtracted.

- First find the common denominator of the real numbers as described in the previous chapter.

- Find the common denominator of the literal part by applying the rules above for finding the LCM.

- After the above steps, add or subtract the numerators.

## PRACTICAL COMMENTS

- Fractions containing real and literal numbers are common in electronics. A good example is the formula for finding the capacitive reactance of a capacitor when applied in an AC circuit. Capacitive reactance is the opposition a capacitor offers to the flow of AC current. It is expressed in ohms and the formula is $X_C = \frac{1}{2\pi fC}$

  The real numbers are 1 and $2\pi$ ($\pi$ is a constant value real number). The literal numbers are f and C. There are many other examples that you will come across in your study of electronics.

## STUDENT QUIZ

Name ______________________________________________

1. The prime factors of a literal number such as $b^2$ is b•b.
   a. true
   b. false
2. To factor a number with literal terms, simply add all the terms together.
   a. true
   b. false
3. To multiply monomials, multiply the real parts and then the literal parts.
   a. true
   b. false
4. Division of literal numbers sometimes involves cancelling of the same literal numbers.
   a. true
   b. false
5. It is not necessary to find a common denominator when adding fractions with different denominators.
   a. true
   b. false
6. Determine the prime factors of $36a^2b^2$.
   a. 2 2 3 3 a b
   b. 2 3 a a b b
   c. 2 3 3 3 a a b b
   d. 2 2 3 3 a a b b
7. Find the LCM of the following numbers. $a^3bc$, $ab^2c$
   a. $a^2b^2c$
   b. $a^2bc$
   c. $a^2b^2c$
   d. $a^3b^2bc$
8. Multiply the following. $3a^2b^2 \bullet 4ab^{-1}c$
   a. $12a^2b^2c$
   b. $12a^3bc$
   c. $12a^3b^2c$
   d. 12abc
9. Reduce the following to the lowest terms. $84a^3bc^2/12a^bc$
   a. 7abc
   b. $7a^2bc$
   c. $7a^2c$
   d. $7abc^2$
10. Multiply the following and reduce to the lowest terms. $(6b/11c) \bullet (4b/3c^2)$
    a. $4b^3/11c^3$
    b. $24b^2/33c^2$
    c. 4b/11c
    d. 6bc

11. Divide the following. $(3a/9) \div (9a/18)$
    a. a/6
    b. 2/3
    c. 18a
    d. 2/3a
12. Perform the following and reduce to the lowest terms.
    (3ab/11) + (9ab/11) - (4ab/11)
    a. 8ab/33
    b. 8/11
    c. 8ab/11
    d. 8b/11a
13. Perform the following and reduce to the lowest terms. 11/9a - 5/9a.
    a. 11/9a
    b. 6/a
    c. 11/45a
    d. 2/3a
14. Perform the following and reduce to the lowest terms. 3b/4 + 3b/7
    a. 33b/28
    b. 3b/28
    c. 6b/28
    d. 12b
15. Perform the following and reduce to the lowest terms. 4/3b - 7/3b
    a. -3/3b
    b. -1/b
    c. 1/b
    d. 3/7b

# LINEAR EQUATIONS

CHAPTER 8

## IDENTIFYING EQUATIONS

- An ***equation*** is an algebraic expression separated by an equal sign (=).
- The equation says that the left side value must equal the right side value.
- An equation where the unknown is raised to the first power is called a ***linear equation***.
- An equation with a power such as $a^2$ is known as a ***second degree equation***.

## LINEAR EQUATIONS

- Solving for an unknown in a linear equation is simple if you remember that you can do anything to one side of an equation as long as you do the same to the other side. This maintains the equality.
- Any mathematical operation can be performed on an equation as long as the equality is maintained.

## SECOND-DEGREE EQUATIONS

- A ***second-degree*** equation has at least one second power exponent.
- Second-degree equations can involve taking a square root of a number with an exponent of 2.
- The square root of a square of a number is that number.
- The square of the square root of a number is that number.

## APPLICATIONS

- The field of electronics uses many equations to express mathematically how a circuit operates.

- A basic equation will state an electronic principal. It is necessary to be able to solve for any unknown in an equation. An example is Ohm's Law. The basic equation is $I = \frac{V}{R}$. The other forms are $V = IR$ and $R = \frac{V}{I}$.

## PRACTICAL COMMENTS

- To achieve success in your study of electronics it will be necessary for you to be able to manipulate equations. Our example of the Ohm's Law manipulation above is a case in point. Of course, you will use the Ohm's Law equations so often that you will have them memorized. There will be many formulas that you will not use often. Under these conditions, it is far easier to remember the basic formula, and then if you need to solve a variation of the formula for a different unknown, manipulate the equation so you can solve it.

## STUDENT QUIZ

Name ______________________________________________

1. The sign (=) in an equation means that each side of the equation must have an equal value.
   a. true
   b. false
2. The linear equation will have unknowns raised to a power of one only.
   a. true
   b. false
3. An equation such as $x^3 = 81$ is a second-order equation.
   a. true
   b. false
4. If you add a 7 to one side of an equation, you must subtract a 7 from the other side to maintain the equality.
   a. true
   b. false
5. This equation $(x^2)^{1/2} = x$ is
   a. true
   b. false
6. This equation $3b + 7 = 9$ is known as
   a. second-degree equation.
   b. a third-degree equation.
   c. a forth-degree equation.
   d. a linear equation.
7. In this equation, $a/3 = 4$
   a. $a = 1\ 1/4$
   b. $a = 12$
   c. $a = 5/4$
   d. $a = 1\ 3/4$
8. Solve for the unknown in this equation. Round to 3 decimal places. $8/a^2 = 4$
   a. 0.707
   b. 2
   c. 1.414
   d. 4
9. Solve for the unknown in this equation. $X^2 - 4 = 32$
   a. $x = 1/6$
   b. $x = 6$
   c. $x = 0.167$
   d. $x = 3$
10. Solve $f_c = 1/2\pi RC$ for C.
   a. $C = 1/2\pi R$
   b. $C = 2\pi f_c R$
   c. $C = 1/2\pi R f_c$
   d. $C = 2\pi / f_c R$

11. Solve $\beta = I_C/I_B$ for $I_B$.
   a. $I_B = I_C/\beta$
   b. $I_B = \beta I_C$
   c. $I_B = \beta/I_C$
   d. $I_B = I_B + I_C$
12. The formula for finding power is $P = I_2R$. If $P = 10$ mW and $R = 10$ kΩ, find I.
   a. 10 A
   b. 10 mA
   c. 1 mA
   d. 0.1 mA
13. The formula for finding the resistance of two resistors in series is $R_T = R_1 + R_2$. If $R_T = 4.7$ kΩ and $R_2 = 3.3$ kΩ, find $R_1$.
   a. 10 kΩ
   b. 8 kΩ
   c. 1.94 kΩ
   d. 1.4 kΩ
14. A formula for finding the total resistance of two resistors in parallel is $R_T = (R_1R_2)/(R_1 + R_2)$. If $R_T = 68$ kΩ and $R_1 = 100$ kΩ, find $R_2$.
   a. 4.7 Ω
   b. 32 kΩ
   c. 212.5 kΩ
   d. 168 kΩ
15. The potential difference across the forward biased pn junction of a transistor is $V_{BE} = V_B - V_E$. If $V_{BE} = -0.7$ V and $V_E = -16$ V, find $V_B$.
   a. -16.7 V
   b. 16.7 V
   c. -17.7 V
   d. 17.7 V

# FACTORING ALGEBRAIC EXPRESSIONS

CHAPTER 9

## MULTIPLICATION OF POLYNOMIALS BY MONOMIALS

- One way of multiplying a polynomials by a monomial is to multiply each term of the polynomial by the monomial. An example of this is the following, $4a(3a + 5) = 12a^2 + 20$.
- Often like terms show up. If so, just combine them. If the result of a multiplication is $2ab - 4ab + 9ab$, then combine them into $7ab$.
- If a fraction is a part of the polynomial, multiply the numerator and reduce if possible.
- It is common practice to list your answer in descending order of the powers, such as $x^3 - x^2 - x$.

## MULTIPLICATION OF BINOMIALS BY BINOMIALS

- Multiplying two binomials together requires that each term of one binomial be multiplied by each term of the other.
- An expression containing four terms will result. Sometimes a combining of terms may be necessary.
- The square of a binomial is the same as multiplying two like binominals. For example, $(a + 2)^2 = (a + 2)(a + 2) = a^2 + 4a + 4$. Notice that the like terms are combined.

## DIVISION OF POLYNOMIALS

- Each term in the numerator must be divided by each term in the denominator.
- An example of this is $(a^2 - 3a + 12)/a = a - 3 + 4/a$.
- A method of long division can also be used.

## FACTORING POLYNOMIALS

- Factoring polynomials is the opposite of multiplying polynomials by a monomial.

- Any group of terms can be factored by any term that is common to all of the terms. Take the term $3a^2 + 9a$. The term common to all of these terms is 3a. Dividing each of these terms by 3a gives a result of $3a(a + 3)$.

## FACTORS OF TRINOMIALS

- Factoring a trinomial requires an inspection of the real numbers to find two numbers whose product equals the last term of the trinomial and whose sum equals the second term.

- An example of this type of factoring is the term $x^2 + 11x + 30$ when factored results in $(x + 5)(x + 6)$. Note the sum of 5 and 6 is 11 and the product is 30.

- Negative terms must be watched for the correct sign.

# PRACTICAL COMMENTS

- In your study of electronics, factoring is a most important facet. This is especially true when studying and working with the mathematics of digital circuits. Simplifying digital circuits uses the factoring process. For example, a digital logic circuit might give you an expression $AB + B$. Using the factoring principles, this factors to $B(A + 1)$. This may not seem simpler but you will learn that in logic circuits $(A + 1) = 1$. This means that the circuit simplifies to B. You will learn to use this factoring in digital logic.

## STUDENT QUIZ

Name ______________________________

1. To multiply a polynomial by a monomial means adding each term of the polynomial and multiplying the sum by the monomial.
   a. true
   b. false
2. If like terms appear as a result of a multiplication, it is usually a good idea to combine them.
   a. true
   b. false
3. Multiplying two binomials usually results in an expression of four terms. These may be combined if possible.
   a. true
   b. false
4. Division of a polynomial by a monomial means to divide each term by the monomial.
   a. true
   b. false
5. Factoring of a polynomial is a very useful mathematical tool in electronics.
   a. true
   b. false
6. Perform the indicated operation and combine any like terms. $5(x + 7)$
   a. $5x + 2$
   b. $5x - 2$
   c. $5x - 35$
   d. $5x + 35$
7. Perform the indicated operation and combine any like terms.
   $6a(a^2 + 4a + 6) + 2a(4a + 12)$
   a. $6a^2 + 32a + 60$
   b. $a^3 + a^2 + a$
   c. $6a^3 + 32a^2 + 60a$
   d. $a + 24a^2 + 12$
8. Perform the indicated operation and combine any like terms.
   $2a(3a + 6 + 2a^2/3)$
   a. $6a^2 + 3a + 4a^3/3$
   b. $6a^2 + 12a + 4a^3/3$
   c. $3a^2 + 6a + 4a^3/3$
   d. $3a^2 + 9a + 4a^3/3$
9. Perform the indicated operation and combine any like terms.
   $(a - 3)(a + 7)$
   a. $a^2 + 4a - 21$
   b. $3a^2 - 4a + 21$
   c. $21a^2 + 4a - 21$
   d. $a^2 - 4a + 21$

10. Perform the indicated operation and combine any like terms.
    $(x + y)(2x - 3y)$
    a. $x^2 - xy + 3y^2$
    b. $2x^2 + xy + 3y^2$
    c. $2x^2 - xy - 3y^2$
    d. $3y^2 + xy - x^2$
11. Perform the indicated operation and combine any like terms.
    $(x^3 - 4x^2 + 3x) \div 2x$
    a. $2x^2 - 4x + 3$
    b. $x^2/2 + 2x - 2/3$
    c. $x^2/2 - 2x + 2/3$
    d. $x/2 - 2 + 2x/3$
12. Perform the indicated operation and combine any like terms.
    $(a^4 - 3a^3 + 6a^2 - a) \div a$
    a. $a^3 + 3a^2 - 6a - 1$
    b. $a^3 - 3a^2 + 6a - 1$
    c. $a^3 - a^2 + 6a - 1$
    d. $a^3 + a^2 + 6a + 1$
13. Factor and reduce the following to the lowest terms. $9x - 3$
    a. $(3x - 1)$
    b. $3(9x - 1)$
    c. $3(3x + 1)$
    d. $3(3x - 1)$
14. Factor and reduce the following to the lowest terms.
    $28a^4b^3c^2 - 14a^3b^2c + 7a^2b^2c$
    a. $4a^2bc - 2a + 1$
    b. $4a^2b^2c - 4ab + c$
    c. $4a^2bc^2 - 4bc + 1$
    d. $4abc - 4bc + 1$
15. Factor and reduce the following to the lowest terms.
    $x^2 + 2x - 63$
    a. $(x + 7)(x - 9)$
    b. $(x - 7)(x + 9)$
    c. $(x + 7)(x - 16)$
    d. $(x - 7)(x + 12)$

# FRACTIONAL EQUATIONS

# CHAPTER 10

## GENERAL EQUATIONS

- Polynomial equations often come with some or all of the terms in the form of a fraction.
- Equations with fractions must be treated as any other fraction. Find the common denominator (CD) and multiply all terms on both sides of the equation by the CD.
- Solve the resulting equation as before.
- An example is $1/3 + x/4 = 3$. The CD is 12. Multiply each term $12/3 + 12x/4 = 36$. This then reduces to $4 + 3x = 36$. From this we get $x = 32/11$.

## SOME REAL EQUATIONS

- A common equation used in electronics is $R_T = \dfrac{R_1 R_2}{R_1 + R_2}$ . This gives the value of total resistance of two resistors in parallel. This equation is in the form of a fraction.
- Another common equation is $A_V = \dfrac{R_f}{R_{in}} + 1$ . This equation is for the voltage gain of a non-inverting op-amp amplifier. It too is in the form of a fraction.
- These practical formulas can often be solved by the use of an electronic calculator.

## QUADRATIC EQUATIONS

- The quadratic equation most used in electronics is in the form of $ax^2 + bx + c = 0$.

- A standard formula for solving these types of equations is $x = -b \pm \frac{\sqrt{b^2 - 4ac}}{2a}$ .

- If the quadratic equation can be factored, factor it into the terms. Set each term to 0, and solve for each value of x.

- The equation can also be solved by placing the equation in the standard quadratic form and solving for the unknown from the above formula.

## PRACTICAL COMMENTS

- As mentioned in a previous chapter, it is far easier to know how to manipulate a standard formula to find the various unknowns, than to remember all of the variations of each formula you will encounter.

## STUDENT QUIZ

Name ______________________________________________

1. Polynomial equations will not be in fraction form.
   a. true
   b. false
2. If a polynomial equation does come in fraction form, then a common denominator must be found.
   a. true
   b. false
3. Electronic formulas often result in fractional equations.
   a. true
   b. false
4. A quadratic equation may be factored and each factor set to zero and solved.
   a. true
   b. false
5. Factoring is the only method available to solve quadratic equations.
   a. true
   b. false
6. Solve the following for the unknown, $2x/7 + 4 = 6$
   a. 14
   b. 42
   c. 7
   d. 3.5
7. Solve the following for for the unknown, $R/4 - 2 = 2R/3$
   a. 5
   b. -5
   c. -24/5
   d. -5/24
8. Solve the following for $R_2$, $f = 1/2\pi C(R_1 + R_2)$
   a. $R_2 = (1/2\pi fC) + 1$
   b. $R_2 = (1/2\pi f) - C$
   c. $R_2 = C + 2\pi f$
   d. $R_2 = (1/2\pi fC) - R_1$
9. Solve the following for the unknown, $x^2 - x - 56 = 0$
   a. $x = -7, -8$
   b. $x = -7, 8$
   c. $x = 7, 8$
   d. $x = 56$
10. Solve the following for the unknown, $x^2 - 24x - 20 = 0$
    a. $x = 24.81, -0.805$
    b. $x = -24.81, 0.805$
    c. $x = -24.81, -0.805$
    d. $x = 24.81, 0.805$

# DC CIRCUIT ANALYSIS: KIRCHHOFF'S LAWS

CHAPTER 11

## KIRCHHOFF'S CURRENT LAW

- The law known as ***Kirchhoff's current law*** states that the current that flows into a junction must equal the current that leaves the junction.
- Mathematically this can be stated as an equation, $I_{IN} = I_{OUT}$.
- Expressed another way $I_{IN} - I_{OUT} = 0$
- Any circuit consists of three basic parts: a source, a path, and a load. The source might be a battery. The path is a piece of wire (a conductor), and the load could be a resistor or a lamp.
- A ***series circuit*** has only one current. The current leaving the source must equal the current entering the source.
- In a ***parallel circuit*** there is more than one path for the current to take in its way back to the source. Each junction point for the current will produce another parallel path.
- The current into a junction equals the current leaving it.
- A ***series-parallel or combination circuit*** has various combinations of series and parallel loads.

## KIRCHHOFF'S VOLTAGE LAW

- ***Kirchhoff's voltage law*** states that the sum of the potential drops around any path must equal 0. For two drops this would be expressed as $V_1 + V_2 - V_T = 0$
- Stated another way the sum of the voltage drops around any closed loop must equal the source voltage. $V_T = V_1 + V_2 + \ldots V_n$

## POLARITY

- With electron current flow the end of a load where the current enters a negative is assigned a negative sign and the other end is positive.

- Conventional current flow is just the opposite.

- Observe the proper polarity in going around the loop. Voltae drops are usually considered positive, while a source is negative, indicating an increase of potential. Conventional current flow is the opposite.

- A loop may contain more than one voltage source.

## PRACTICAL COMMENTS

- There are two schools of thought on current flow. One group sees the current flowing from the positive terminal of the source back to the negative side. This is called conventional current flow. The opposite school sees the current flowing in the direction that the electrons must take, from the negative terminal towards the positive. This is called electron current flow. In actuality either method can be used (and are used) to analyze current flow correctly.

- Kirchhoff's Voltage Law can be used all of the time to analyze circuits. In a series circuit, for example, the sum of the voltage drops must equal the source. This one fact will enable you to troubleshoot many electronic circuits.

## STUDENT QUIZ

Name ____________________________________________

1. Kirchhoff's current law states that the sum of the currents leaving a junction must be approximately equal to the currents entering that junction.
   a. true
   b. false
2. The sum of the voltage drops in any loop must equal zero.
   a. true
   b. false
3. There are many currents in a series circuit.
   a. true
   b. false
4. The expression $I_T = I_1 = I_2$ is correct to express the currents in a parallel circuit.
   a. true
   b. false
5. A series-parallel circuit can have many series and parallel loads.
   a. true
   b. false

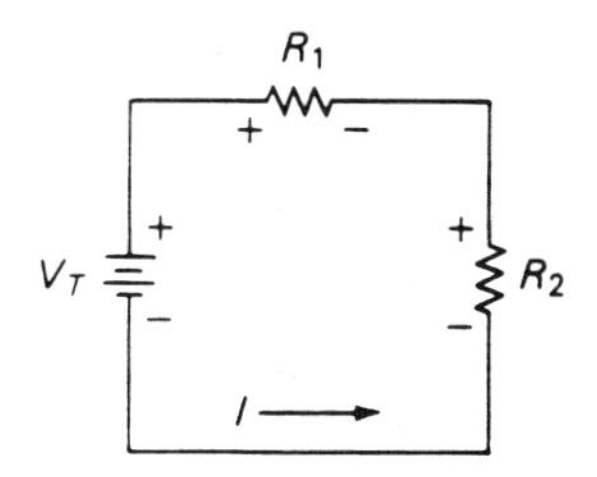


**Figure 11-1**

6. See figure 11-1. If $V_T$ = 17 V and $V_2$ = 12 V, find $V_1$.
   a. 29 V
   b. 12 V
   c. 17 V
   d. 5 V
7. See figure 11-1. If the current through $R_2$ = 15 mA, the current through $R_1$ is __________ and the total current is __________.
   a. 17 mA, 12 mA
   b. 12 mA, 17 mA
   c. 15 mA, 15 mA
   d. 30 mA, 18 mA

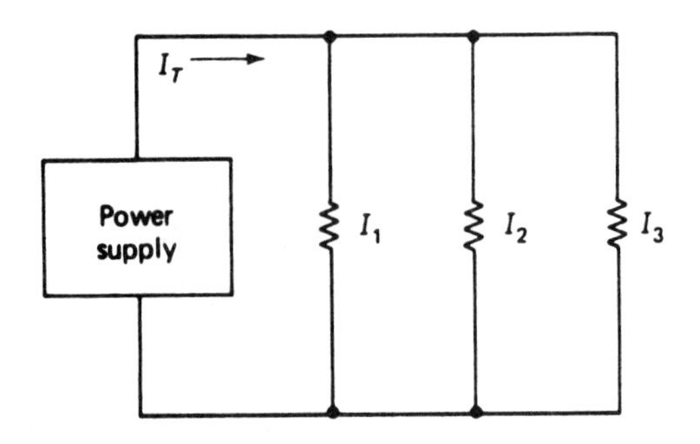


**Figure 11-2**

8. See figure 11-2. Find $I_T$ if $I_1$ = 33 mA, $I_2$ = 16 mA, and $I_3$ = 12 mA.
   a. 61 mA
   b. 20.3 mA
   c. 45 mA
   d. 67 mA

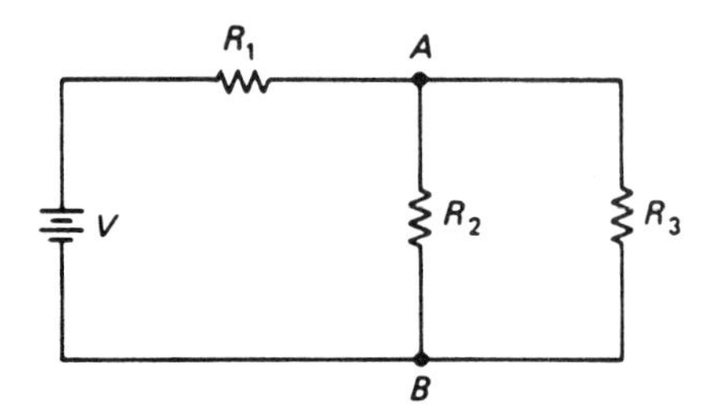


**Figure 11-3**

9. See figure 11-3. If V = 50 V, $V_1$ = 22 V, find $V_2$.
   a. 50 V
   b. 22 V
   c. 72 V
   d. 28 V
10. See figure 11-3. If $V_1$ = 250 mV and $V_2$ = 1.7 V, find V.
   a. 250 mV
   b. 1.45 V
   c. 1.95 V
   d. -1.45 V

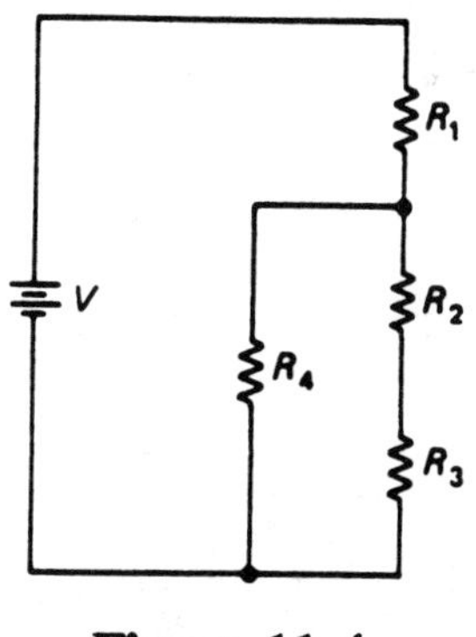


**Figure 11-4**

11. See figure 11-4. Which of the following expressions is correct?
    a. $V = V_4 + V_2 + V_1$
    b. $V_4 = V - V_1 + V_2$
    c. $V_1 = V - V_4$
    d. $V_2 = V_1 + V_3$
12. See figure 11-4. Which of the following statements is correct?
    a. $R_1$ is in series with $R_2$.
    b. $R_1$ is in parallel with $R_4$.
    c. $R_2$ is in parallel with $R_3$.
    d. $R_2$ is in series with $R_3$.

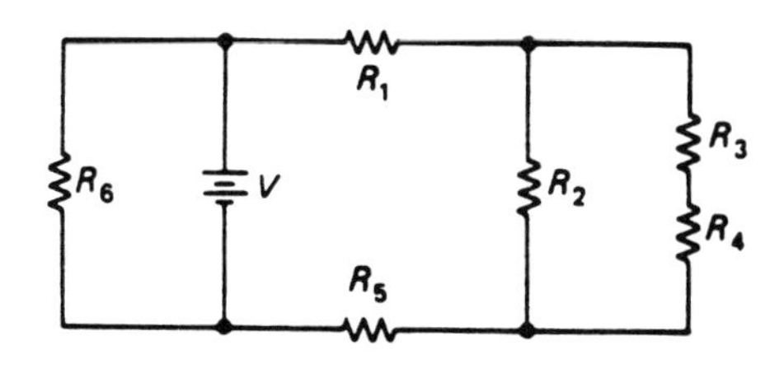


**Figure 11-5**

13. See figure 11-5. If $V = 24$ V, $V_1 = 3$ V, $V_5 = 2$ V, and $V_4 = 1$ V, find $V_3$.
    a. 18 V
    b. 20 V
    c. 21 V
    d. 23 V

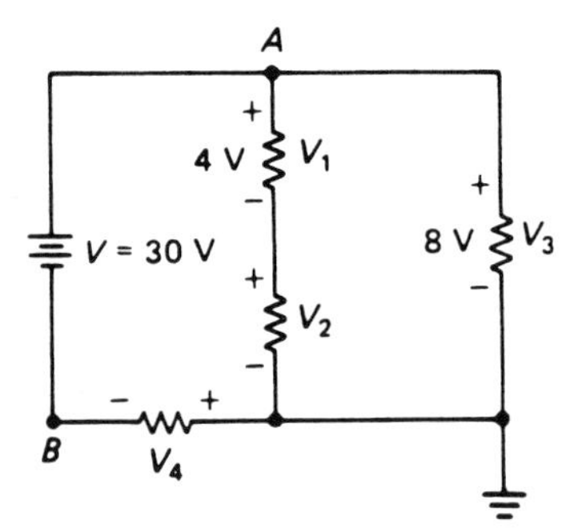


**Figure 11-6**

14. See figure 11-6. Find the voltage from point A to ground.
    a. 3 V
    b. 5 V
    c. 8 V
    d. 4 V
15. See figure 11-6. Find the voltage from point B to ground.
    a. 4 V
    b. 8 V
    c. 22 V
    d. 16 V

# DC CIRCUIT ANALYSIS: OHM'S LAW

# CHAPTER 12

## CIRCUIT RESISTANCE

- In a series circuit, the total resistance is equal to the sum of the individual resistances. $R_T = R_1 + R_2 + \ldots R_N$.

- ***Conductance (G)*** is the reciprocal of resistance. G = 1/R

- In a parallel circuit the total resistance is less than any of the branch resistances.

- The formula for finding the total resistance of resistors in parallel is

$$R_T = \frac{1}{\frac{1}{R_1} + \frac{1}{R_2} + \frac{1}{R_3} + \ldots + \frac{1}{R_N}} \quad \text{or } R_T = \frac{1}{G_T} = \frac{1}{G_1 + G_2 + G_3 + \ldots + G_N}$$

- A series-parallel circuit must be analyzed by switching from series laws to parallel laws as required.

- Two resistors in parallel often use the symbol || to indicate that they are in parallel. For example, $R_1 \| R_2$.

## SERIES CIRCUITS

- There is only one current in a series circuit.

- The basic form of Ohm's Law is I = V/R.

- The voltage drops across any load in a series circuit can be found by Ohm's Law. Be sure to use the values for that load if you need the drop across a particular resistor.

- The use of total voltage and total current will give you total resistance.

- Ohm's Law tells us that in any circuit, current and resistance are inversely proportional. If the resistance increases, the current decreases.

- Voltage and resistance are directly proportional, that is, the larger resistance will have the larger voltage drop across it.

- An increase in voltage will cause an increased amount of current to flow. Voltage and current are directly proportional.

**PARALLEL CIRCUITS**

- Ohm's Law applies to parallel circuits as well as all other circuits.

- In a parallel circuit, the voltage is constant across common branches.

- The current flow is greater in the branch with the least resistance.

- Branch currents add in a parallel circuits. $I_T = I_1 + I_2 + \ldots I_N$

**SERIES-PARALLEL CIRCUITS**

- Analysis of these series-parallel circuits involves changing from working with series circuit laws to parallel laws as often as necessary.

- If a branch has series resistances, combine them.

- If the above branch is in parallel with another resistor, then use the parallel rules to solve for the total resistance or current as required.

## PRACTICAL COMMENTS

- Always remember to apply Ohm's Law to an individual resistor using that resistor's values. If you are considering the total circuit, then use the total values of I, V, and R. Mixing these values will cause you to get many incorrect answers.

- An excellent check on your calculations for the total resistance of a parallel circuit is that $R_T$ must always be less than the resistance value of the smallest resistor in the parallel circuit. Of course this is true, because as you add parallel branches the current increases and therefore the resistance must decrease.

- Solving series-parallel circuits can be very easy. Just remember to draw the current paths on your diagram. These paths will show you the series combinations that can be made. Also, when you come to a branch, a parallel circuit must be present. Switch from series to parallel rules as often as needed in any order necessary.

## STUDENT QUIZ

Name ______________________________________________

1. There is only one current in a series circuit.
   a. true
   b. false
2. The total resistance of resistors in parallel is found by $R_T = R_1 + R_2$.
   a. true
   b. false
3. This expression is correct for finding the resistance of resistors in series.
   $R_T = R_1R_2/(R_1 + R_2)$
   a. true
   b. false
4. Loads in parallel share the same voltage.
   a. true
   b. false
5. Ohms's Law applies to all circuits.
   a. true
   b. false
6. Which of the following expressions is *NOT* correct?
   a. V = I/R
   b. I = V/R
   c. R = V/I
   d. V = IR

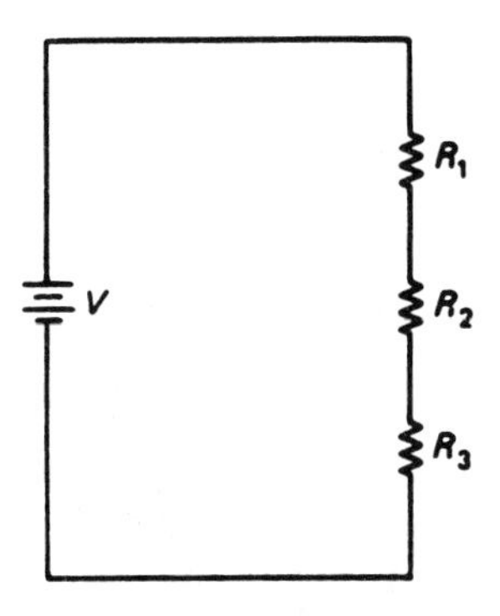


**Figure 12-1**

7. See figure 12-1. If V = 30 V and the three resistors equal 10 kΩ each, find the voltage across $R_2$.
   a. 30 V
   b. 3 V
   c. 12 V
   d. 10 V

8. See figure 12-1. If I = 10 mA, $R_1$ = 5 kΩ, $R_2$ = 3 kΩ and $R_3$ = 4 kΩ, find V.
   a. 50 V
   b. 30 V
   c. 40 V
   d. 120 V

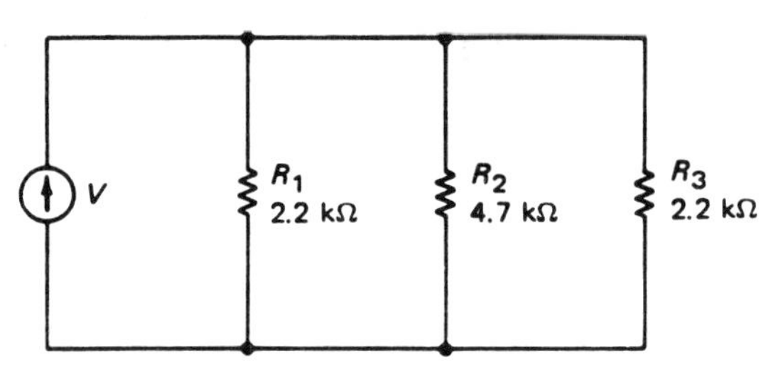


**Figure 12-2**

9. See figure 12-2. Find $R_T$.
   a. 18.6 kΩ
   b. 13.9 kΩ
   c. 0.89 kΩ
   d. 0.47 kΩ
10. See figure 12-2. If $I_3$ = 5 mA, find V.
   a. 11 V
   b. 6.55 V
   c. 93 V
   d. 12 V

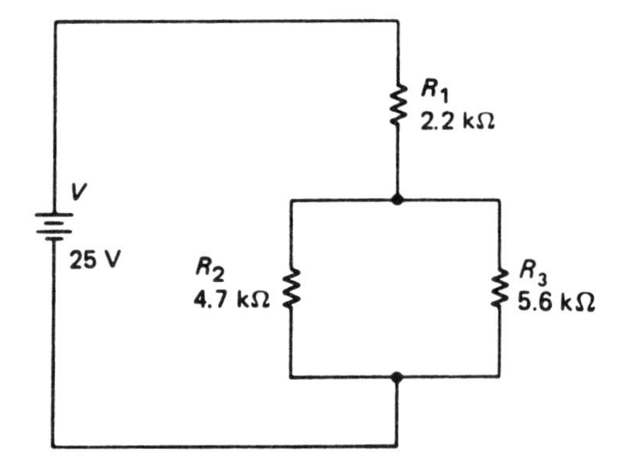


**Figure 12-3**

11. See figure 12-3. Find $I_T$.
    a. 6.98 mA
    b. 5.26 mA
    c. 11.4 mA
    d. 11.6 mA
12. See figure 12-3. Find $R_T$.
    a. 4.7 kΩ
    b. 2.2 kΩ
    c. 10.3 kΩ
    d. 4.75 kΩ
13. See figure 12-3. Find $I_2$.
    a. 2.60 mA
    b. 2.86 mA
    c. 5.69 mA
    d. 9.75 mA

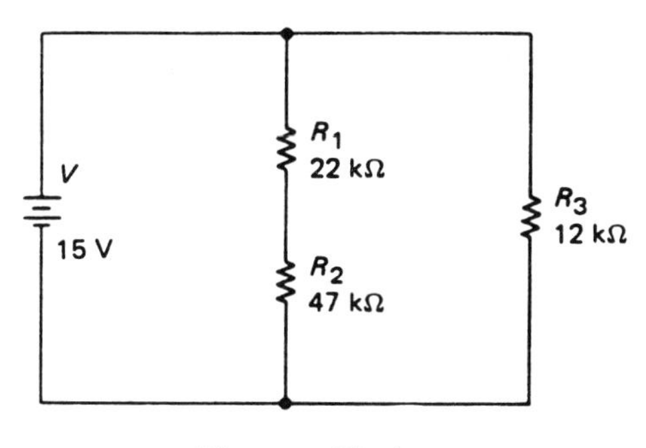


Figure 12-4

14. See figure 12-4. Find $R_T$.
    a. 10.22 kΩ
    b. 5.75 kΩ
    c. 81 kΩ
    d. 26.99 kΩ
15. See figure 12-4. Find $V_2$.
    a. 4.78 V
    b. 15 V
    c. 10.22 V
    d. 12 V

# DC CIRCUIT ANALYSIS: CIRCUIT THEOREMS

# CHAPTER 13

## SUPERPOSITION THEOREM

- The ***superposition theorem*** is a theorem that is used to analyze a circuit that usually contains multiple voltage sources.

- This theorem uses a method of shorting each voltage source except one. Then the calculations are made for the current in a particular load.

- Each voltage is shorted out and the load current is calculated until all sources have been dealt with in a like manner.

- The resultant current through the load is then the algebraic sum of the currents found by the above calculations.

- The polarity of the currents must be watched carefully.

- When the load current has been found, the load voltage can then be calculated.

## THÉVENIN'S THEOREM

- ***Thévenin's theorem*** states that any circuit, no matter how complex, can be represented by one series resistance and one voltage source.

- The ***voltage source*** is called $V_{OC}$.

- The equivalent resistance is labeled $R_{TH}$.

- There are certain steps to take when applying Thévenin's Theorem. The first is to remove the load and calculate the voltage drop across the terminals of the open load. This will be called $V_{OC}$.

- Next, replace the voltage source with a short and calculate the resistance seen at the open circuit terminals. This is $R_{TH}$.

- Last, connect the load back to the output terminals of the equivalent circuit just calculated. Calculate the load current $I_L$ by $I_L = \dfrac{V_{OC}}{R_{TH} + R_L}$

- Calculate the load voltage as follows: $V_L = \frac{V_{OC} R_L}{R_{TH} + R_L}$

## NORTON'S THEOREM

- ***Norton's theorem*** states that any circuit can be represented by a single current source and a parallel resistance.

- To operate Norton's theorem, replace the load with a short circuit and calculate the current through the short. This is $I_{SC}$.

- Next, open the load short and short the voltage source. Determine the conductance at the open circuit terminals. Call this $G_N$.

- Often it is easier to find $R_{TH}$ and then take the reciprocal to find $G_N$.

- Calculate the value of $V_L = \frac{I_{SC}}{G_N + G_L}$.

- The load current can then be found by $I_L = \frac{I_{SC} G_L}{G_L + G_N}$.

# PRACTICAL COMMENTS

- The superposition theorem is very useful for finding the currents and voltages across a load when multiple voltage sources are used. Many transistor circuits are analyzed by this method because there are several DC voltage sources present. The theorem is handy, learn it and use it.

- Thévenin's theorem is used more than the other circuit theorems. Some types of circuits can be solved easily with this theorem. One such circuit is the bridge circuit. Using ordinary Ohm's law on a bridge circuit is very difficult. These bridge circuits are often found in measuring instruments and some types of transistor amplifiers.

## STUDENT QUIZ

Name ______________________________________________

1. The superposition theorem is very handy for calculating the values in a bridge type circuit.
   a. true
   b. false
2. Thévenin's equivalent resistance is called $R_{TH}$.
   a. true
   b. false
3. Thévenin's theorem states that any circuit can be reduced to one voltage source and one resistance.
   a. true
   b. false
4. Norton's theorem states that any circuit can be reduced to several current sources and one parallel resistance.
   a. true
   b. false
5. All of these circuit theorems require certain steps to use them properly.
   a. true
   b. false

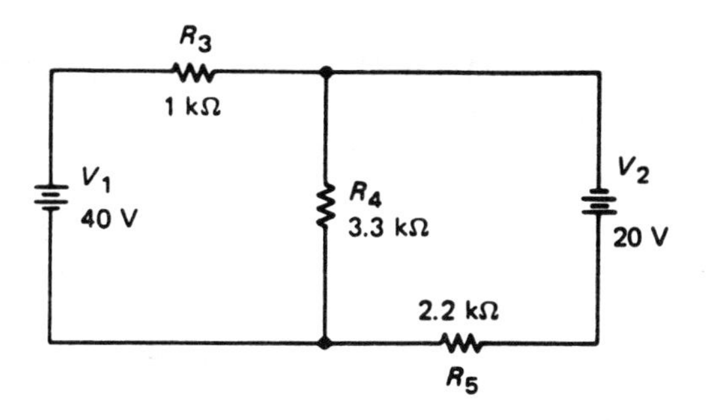


**Figure 13-1**

6. See figure 13-1. Use the superposition theorem and find the value of $V_4$.
   a. 22.76 V
   b. 5.17 V
   c. 17.59 V
   d. 27.97 V
7. See figure 13-1. Use the superposition theorem and find the value of $I_4$.
   a. 17.24 mA
   b. 5.33 mA
   c. 6.74 mA
   d. 22.76 mA

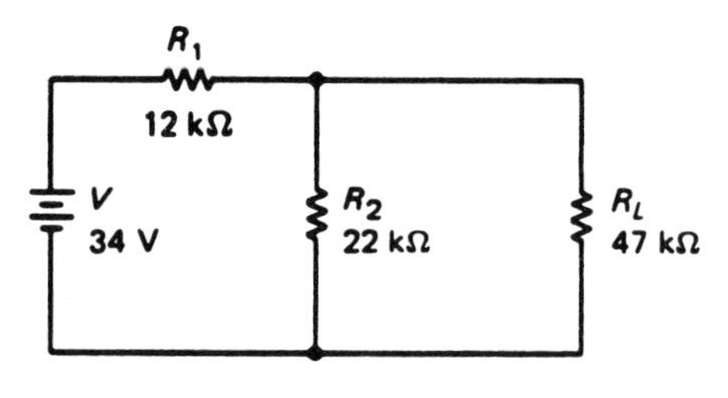


**Figure 13-2**

8. See figure 13-2. Use Thévenin's theorem and find the value of $V_{RL}$.
   a. 15.12 V
   b. 22 V
   c. 1.26 V
   d. 18.88 V
9. See figure 13-2. Use Thévenin's theorem and find the value of $I_{RL}$.
   a. 0.402 mA
   b. 1.26 mA
   c. 0.858 mA
   d. 0.456 mA

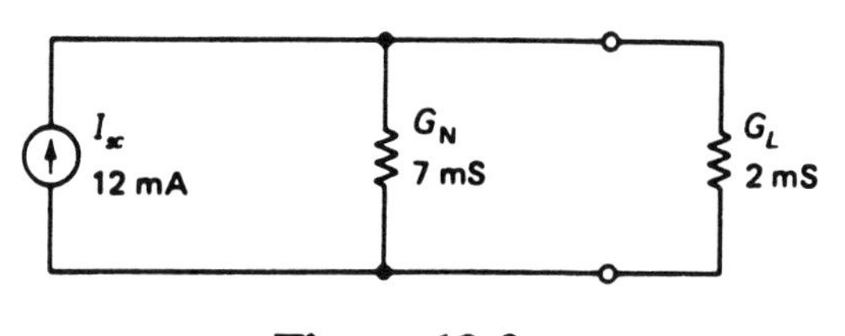


**Figure 13-3**

10. See figure 13-3. Use Norton's theorem to find $I_L$ in this Norton's equivalent circuit.
    a. 26.6 μA
    b. 93.1 μA
    c. 2.66 mA
    d. 9.31 mA

# GRAPHING

CHAPTER

14

## GRAPHING LINEAR EQUATIONS

- A ***linear equation*** is an equation in which each term contains only one variable and no exponents other than one.
- A ***graph*** is a picture of results of an equation as one factor changes.
- A graph of a linear equation must be a straight line.
- A system of ***rectangular coodinates*** is a graph where the horizontal axis (x-axis) is laid out in equal increments. The vertical axis (y-axis) is also linear.
- Positive values for the x-axis are plotted horizontally from the vertical axis towards the right. Negative values are plotted to the left.
- Positive values of y are plotted up and negative values are plotted down.
- The system of coordinates uses a standard notation as in (x, y) to identify a point. The x-coordinate is always first; the y-coordinate is last.
- A coordinate of (2, 3) means that the point is 2 spaces to the right and 3 spaces up.
- The x-axis is called the ***abscissa***. The y-axis is called the ***ordinate***.
- An equation such as $3x - 2y = 5$ will be a straight line, but is is safe to assume that there will be a place where the resulting graph line will Uintercept both the x-and y-axes.
- To find these intercept points, just set x to zero and solve for y. You will have the point coordinate of (0, y). Do the same for $y = 0$. The point coordinate will be (x,0).

## SLOPE OF A LINE

- The ***slope*** of a line is a measure of how fast the line changes. Slope is as the $\frac{\text{change in } y}{\text{change in } x}$ or $\frac{\text{DELTA } y}{\text{DELTA } x}$ or $\frac{\Delta y}{\Delta x}$.

- A slope of 1/3 means that y changes 1 unit while x changes 3.

- We always read the graph so that x is positive.

## SLOPE-INTERCEPT FORM

- A linear equation containing two variables can be written in the ***slope-intercept form***.

- The slope-intercept form is $y = mx + b$.

- In this form, m is the slope and b is the variable.

- Any two-variable linear equation can be rewritten in this form, and the slope and intercept are directly known.

## INTERPRETING GRAPHS

- If a graph is a straight line, then you can find the value of the y-coordinate for any value of x.

- The slope can also be found.

- A graph may not be linear. In this case the slope must be found by finding the tangent at the point on the graph you are interested in.

- There are many curves that are non-linear, and they can also be used to find points or values as one axis changes.

## PLOTTING CURVES

- The curve for the relationship between voltage and current in a circuit with a resistor is linear. The ***independent variable*** is plotted on the horizontal axis. In this case it would be the value for voltage.

- The ***dependent variable*** is plotted on the vertical axis and is current. The current is dependent upon the voltage.

- Look at your data and select an appropriate scale. Then plot your data.

- Linear graphs only require two points to plot. You should use a third point to check for errors.

# PRACTICAL COMMENTS

- Graphs are an important facet in your study of electronics. For example, a graph of the frequency response of your stereo amplifier will show how well it performs at low or high frequencies. This kind of curve is often included in the manufacturer's data on the unit.

- Another vital use of graphing principles is the trace on an oscilloscope. This trace is an instantaneous graph of the way a voltage is changing in a circuit under test. The oscilloscope has a horizontal axis of time, called the timebase. The vertical axis is the voltage present in the circuit. The resulting graph shows the varying of voltage as time changes. This important tool for electronic measurments is used frequently.

## STUDENT QUIZ

Name ______________________________

1. A linear equation contains only one variable in a term.
   a. true
   b. false
2. The coordinate of a point is expressed (y, x).
   a. true
   b. false
3. An intercept is when one variable of an equation equals zero.
   a. true
   b. false
4. The slope of a line equals (delta y)/(delta x).
   a. true
   b. false
5. In an equation in the form of $y = mx + b$, m is the slope and b is the intercept.
   a. true
   b. false

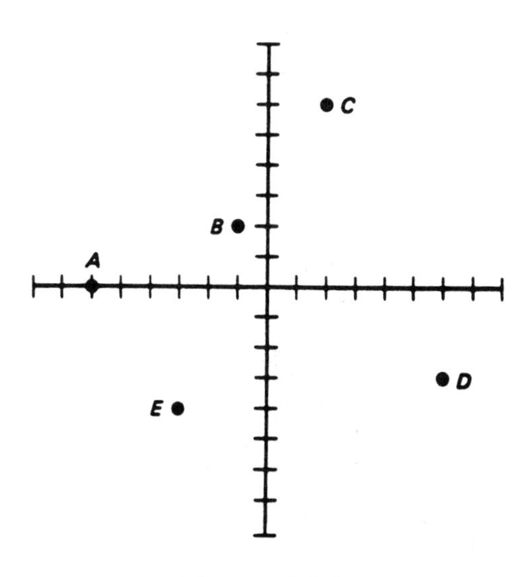


**Figure 14-1**

6. See figure 14-1. The coordinates of point A are
   a. (-1, 7)
   b. (2, 6)
   c. (-6, 0)
   d. 6, -3)

7. See figure 14-1. The coordinates of point D are
   a. (-1, 7)
   b. (2, 6)
   c. (-6, 0)
   d. 6, -3)
8. See figure 14-1. A line drawn between points C and E would have an x-intercept of
   a. (0, -2)
   b. (0, 2)
   c. (-1, 0)
   d. (0, -1)
9. See figure 14-1. A line drawn between points C and E would have a y-intercept of
   a. (0, -2)
   b. (0, 2)
   c. (-1, 0)
   d. (0, -1)
10. See Figure 14-1. Given the equation $y = 2x + 2$. Select two coordinate points from the figure that will fall on the graph of this equation.
    a. C, D
    b. C, E
    c. A, D
    d. A, E
11. See Figure 14-1. Given the equation $y = -x/4 + 3/3$. Select two coordinate points from the figure that will fall on the graph of this equation.
    a. C, D
    b. C, E
    c. A, D
    d. A, E

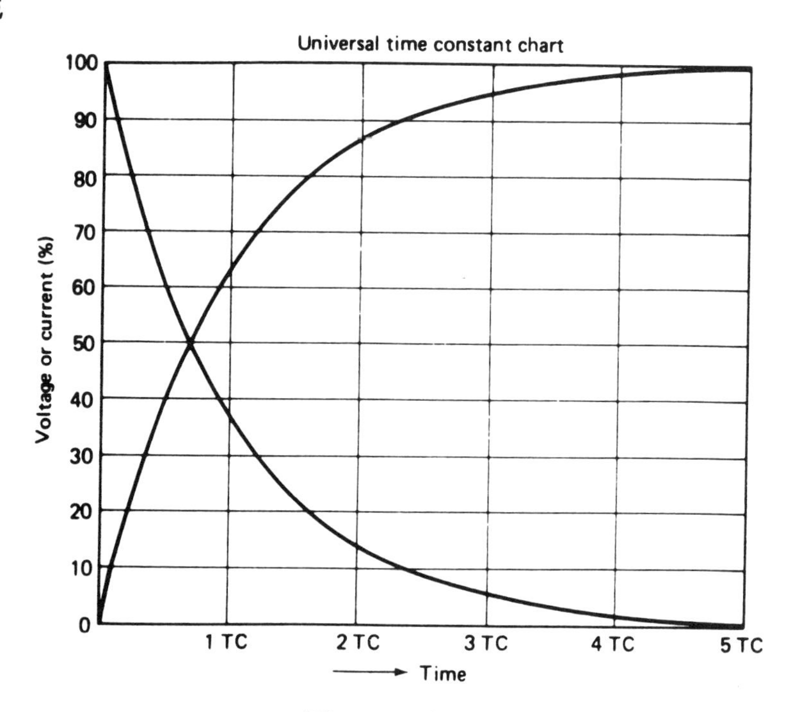


**Figure 14-2**

12. See Figure 14-2. If the applied voltage is 100 V, find the value of $V_C$ at 0.707 TC.
    a. about 63 V
    b. about 50 V
    c. about 23 V
    d. about 16 V
13. See Figure 14-2. If the applied voltage is 22 V, find the number of time constants to charge the capacitor to 22 V.
    a. 1
    b. 2
    c. 3
    d. 4
    e. 5
14. See Figure 14-2. If the applied voltage is 25 V in an RC circuit, find the value of $V_R$ after 3 TC.
    a. about 15.9 V
    b. about 21.6 V
    c. about 23.5 V
    d. about 1.5 V
15. See Figure 14-2. If the applied voltage is 12 V in an RC circuit, find the number of time constants required to charge the capacitor to 7.62 V.
    a. about 1
    b. about 2
    c. about 3
    d. about 4
    e. about 5

# SIMULTANEOUS LINEAR EQUATIONS

CHAPTER 15

## LINEAR EQUATIONS IN TWO UNKNOWNS: GRAPHICAL SOLUTION

- Every linear equation in two unknowns has an unlimited number of solutions.
- If there are two linear equations that are plotted on the same system of rectangular coordinates, then there will be only one solution. One point will be common to each graph as long as the lines are not parallel or the same.
- Two equations that have one common point for a solution are called ***simultaneous equations***.

## LINEAR EQUATIONS IN TWO UNKNOWNS: ALGEBRAIC SOLUTION

- It is possible to solve simultaneous equations by the ***addition or subtraction method.***
- Using the rules for algebra, manipulate the two equations so that the x or y terms are equal.
- Add or subtract the two equations. Use addition if the term signs are different, subtraction if the signs are the same.
- Solve for the unknown, and then substitute in the other equation to find the other unknown.
- The ***substitution method*** of solving simultaneous equations is also used.
- Solve one of the equations for one unknown in terms of the other.
- Substitute this result in the other equation, and solve for the unknown.

## DETERMINANTS

- Another method to solve simultaneous equations is to use ***determinants***.
- A determinant is an array of numbers.
- Arrange the two linear equations in the following general equations:

- Arrange the two linear equations in the following general equations:

  $a_1x + b_1y = k_1$ and

  $a_2a + b_2y = k_2$

- The a's and b's are the coefficients of the unknowns.

- The k's are the constants.

- The ***determinant of the denominator*** is found by cross-multiplying the coefficients of the unknowns. The general equation will be $a_1b_2 - a_2b_1$.

- The general equation of the numerator will be $k_1b_2 - k_2b_1$ for the unknown x, and $a_1k_2$ $-a_2k_1$ for the unknown y.

- Put the numerator and denominator together and solve for the unknowns.

- Solving three simultaneous equations using determinants is similar to the above method.

- If there are three coefficients in a diagonal row, arrange the coefficients in an array and multiply them.

- Substitute the values for the coefficients and solve for the unknowns x, y, and z.

**APPLICATIONS**

- Simultaneous equations can be used to solve circuits. The results will be the same as we discovered using the circuit theorems.

- An example is using Kirchhoff's voltage law in a parallel circuit. Write the two equations around each loop. If there are two loops, two equations will result.

- Solve these equations as simultaneous equations using one of the methods in this chapter.

- If the circuit under analysis has two sources, you must assume the direction of the currents. If you assume incorrectly, the values will be correct but the polarity will be negative.

## PRACTICAL COMMENTS

- Simultaneous equations can be solved in many ways. In relatively simple electronic circuits, often the simplest method is to use Ohm's and Kirchhoff's laws. Learn to solve these equations with the method that involves the fewest steps. The fewer steps you need for a solution, the greater your chances are for an accurate solution.

- Checking your answer is a smart procedure to learn when solving electronic circuits. Find some method for checking your answer. An example might be to use Ohm's law when you have solved for a current or voltage. Are your values correct? Will they give you the correct value of resistance that you started with? Ohm's law will give you a quick check.

## STUDENT QUIZ

Name ______________________________________________

1. Linear simultaneous equations can not be solved by a graphical method.
   a. true
   b. false
2. Addition and subtraction is a method for solving linear simultaneous equations.
   a. true
   b. false
3. One method of solving simultaneous equations is by use of determinants.
   a. true
   b. false
4. Circuit analysis can never be made using simultaneous equations.
   a. true
   b. false
5. Substitution is a method for solving simultaneous equations.
   a. true
   b. false
6. Given these two equations, $3x + 4y = 8$ and $x + y = 4$. Find the value of x and y that will satisfy both equations. Use the addition or subtraction method.
   a. $x = -4, y = 8$
   b. $x = 4, y = -8$
   c. $x = 8, y = -4$
   d. $x = -8, y = 4$
7. Use the substitution method to solve these two equations.
   $3x + y = 12$ and $2x + 2y = 8$
   a. $x = 4, y = 16$
   b. $x = -4, y = 8$
   c. $x = 2, y = 2$
   d. $x = 4, y = 0$
8. Use determinants to solve the following two equations.
   $4x + 5y = 20$ and $3x + y = 9$
   a. $x = 2\ 3/11, y = 2\ 2/11$
   b. $x = 2, y = 11$
   c. $x = -11, y = -2$
   d. $x = 11, y = 2\ 2/11$

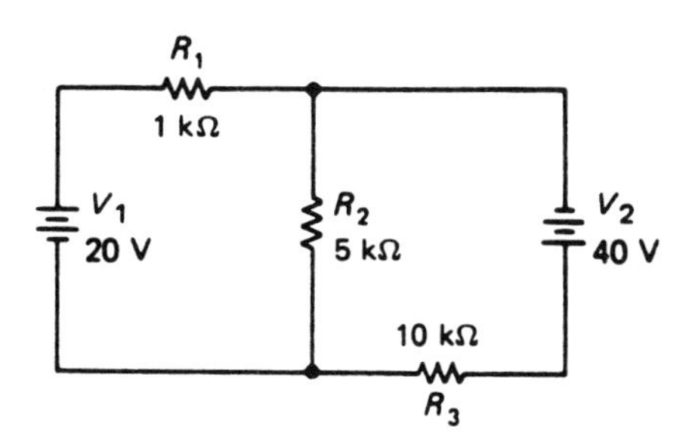


**Figure 15-1**

9. See figure 15-1. Use simultaneous equations to find $V_2$.
   a. 26.67 V
   b. 12.3 V
   c. 16.67 V
   d. 20 V
10. See figure 15-1. Find the value of $I_2$.
   a. 20 mA
   b. 16.67 mA
   c. 2.46 mA
   d. 26.67 mA

# COMPLEX NUMBERS

CHAPTER 16

## IMAGINARY NUMBERS

- The square root of a negative number is an ***imaginary*** number.
- It is usual to signify the square root of a negative number as $(-36)^{1/2} = 6(-1)^{1/2}$. The number 6 is real and the $(-1)^{1/2}$ is the imaginary part.

- In electronics we shorten the imaginary to ***j***. $j = (-1)^{1/2}$.

## COMPLEX NUMBERS

- A ***complex number*** consists of a number with a real and an imaginary part, such as 3 + j6. The 3 is real and the j6 is imaginary.

- Real numbers can be combined. Imaginary numbers can also be combined. An example is 3 + 4 + j3 + j4 = 7 + j7.

- Addition and subtraction of complex numbers consists of adding (or subtracting) the real parts and then doing the same for the imaginary parts.

- For example, (3 + j5) + (6 + j2) = 9 + j7.

- Most scientific calculators will perform these calculations for you.

- To multiply complex numbers, use the rules of algebra. Remember that $j \bullet j = j^2 = -1$.

- The ***conjugate*** of a complex number is that same number with the sign of the imaginary part changed.

- To divide a complex number we multiply the numerator by the conjugate of the denominator.

- Again the use of the calculator is very handy.

- A complex number can be graphed on a system of rectangular coordinates.

- The x-axis shows the j values and the y-axis shows the real values.

- A rectangular number is properly graphed by drawing a rectangle with one corner on a poin such as the origin, $0 + j\,0$, and the opposite corner at the designated number, such as $3 +$ $j\,4$ where the horizontal part of the rectangle is 3 units long and the vertical part of 4 unit high. This rectangle contains 12 square units.

- An arrow (called a vector) can be drawn along the diagonal of the rectangle formed by th rectangular number $a + j\,b$. The starting point of the arrow is at the origin ( $0 + j\,0$ ) an extends directly up to the opposite point, such as $a + j\,b$. This vector is usually given usin polar numbers of the form $c\ \underline{/\Theta}$ where c is the length of the diagonal line and $\Theta$ is the angl which is the arctangent of $b/a$ .

## PRACTICAL COMMENTS

- Other math books use i for imaginary numbers. However in electronics the i is used fc current. The next letter in the alphabet after i is j, so j is used in electronics.

- Imaginary or rectangular numbers are used extensively in the analysis of AC circuits. Th j represents the phase angle $\Theta$ between the voltage and the current of a circuit which contain R or G or C or L. Of course DC circuits do not have a differing phase relationship becaus when the voltage increases, the current increases at the same time. In AC circuits, Ohm Law still works but with a different twist which includes j.

- The answer to any problem with rectangular numbers must be in the form of $\pm\, a \pm j$ where a is clearly separated from b.
  $\frac{-4 + j\,5}{3} = \frac{-4}{3} + j\,\frac{5}{3}$ so $a = \frac{-4}{3}$ and $b = \frac{5}{3}$

- Always replace $j^2$ with -1 as soon as you get it.

- Never leave j in the denominator.
  $\frac{7}{j2} = \frac{7 \times j}{j2 \times j} = \frac{j\,7}{-\,2} = -j\,\frac{7}{2}$

- To get rid of j in a denominator of form $a + j\,b$, multiply the entire fraction by a fractic numerator and denominator equal to the complex conjugate of the denominator: $a - j\,b$.
  $\frac{4 - j\,1}{3 + j\,2} = \frac{(4 - j\,1)(3 - j\,2)}{(3 + j\,2)(3 - j\,2)} = \frac{12 - j8 - j3 + j^2 2}{9 - j^2\,4} = \frac{10}{13} - j\,\frac{11}{13}$

## STUDENT QUIZ

Name ______________________________

1. The solution to $x = (-6)^{1/2}$ is an real number.
   a. true
   b. false
2. The term j is used to signify imaginary numbers.
   a. true
   b. false
3. Imaginary numbers can be combined similar to real numbers.
   a. true
   b. false
4. The graph of a complex number is called a vector.
   a. true
   b. false
5. When it comes to complex numbers, calculators are not useful.
   a. true
   b. false
6. Perform the following. $j^2 \bullet j^3$
   a. j6
   b. -j6
   c. 18j2
   d. -j18
7. Perform the following. (15 + j3) + (-1 + j3)
   a. -15 + j6
   b. 14 - j
   c. 14 + j6
   d. 15 - j6
8. Perform the following. (2 - j3) - (6 + j4)
   a. 4 - j7
   b. -4 + j7
   c. -4 - j7
   d. 4 + j
9. Perform the following. (2 + j3)(3 - j9)
   a. 33 + j9
   b. -33 + j9
   c. 33 - j9
   d. 6 + 18j
10. Perform the following. (3 - j2) ÷ (3 + j2)
    a. 0.923 - j0.385
    b. 0.385 - j0.923
    c. -1.08 + j1.08
    d. -1j

# THE RIGHT TRIANGLE

# CHAPTER 17

## SIDES AND ANGLES

- A ***right triangle*** is a triangle where one angle is 90° or a right angle.
- The longest side of a right triangle is called the ***hypotenuse***.
- The angles are often labeled A, B, and C, where C is the right angle.
- The side opposite to the angle A is called the ***side opposite***.
- The ***adjacent side*** is the side closest to the angle A.
- The angle A is often called ***theta*** ($\theta$) in electronics.
- The angle B is called ***phi*** ($\Phi$).
- The sum of the angles $\theta$ and $\Phi$ is 90°.
- In the electronics field the hypotenuse is often called Z for impedance, the adjacent side is R for resistance, and the opposite side is X for reactance.

## PYTHAGOREAN THEOREM

- The ***Pythagorean theorem*** states that the sum of the squares of the sides equals the square of the hypotenuse or $c^2 = a^2 + b^2$. A more convenient form is $c = \sqrt{a^2 + b^2}$.
- If any two sides of a right triangle are known, then the third side can be found by applying the Pythagorean theorem.
- An AC circuit will apply this theorem as $Z = \sqrt{R^2 + X^2}$.
- The calculator is again very useful for doing this type of calculation.

- The Pythagorean theorem is useful for finding the length of a side of a right triangle where knowing the value of an angle is not required.

- The sides of a right triangle in relationship to one another are discussed in ratios known as ***trigonometric functions***.

- The functions we will work with are called the ***sine***, ***cosine***, and ***tangent***.

- The sine of an angle $= \dfrac{\text{opposite side}}{\text{hypotenuse}}$ or $\sin\theta = \dfrac{o}{h}$
- The cosine of an angle $= \dfrac{\text{adjacent side}}{\text{hypotenuse}}$ or $\cos\theta = \dfrac{a}{h}$
- The tangent of an angle $= \dfrac{\text{opposite side}}{\text{adjacent side}}$ or $\tan\theta = \dfrac{o}{a}$

## TRIGONOMETRIC TABLES

- The three basic trigonometric functions for various angles are listed in a table called a ***trig table***.

- The trig values are more easily found by your calculator.

## TRIGONOMETRIC EQUATIONS

- The trig relationships are actually formulas you can manipulate to obtain other unknowns. For example, $\tan\theta = a/b$ is the basic tangent function. Also $a = b\tan\theta$, and $b = a/\tan\theta$.

- The calculator is again a useful tool.

# PRACTICAL COMMENTS

- A easy way to remember the trig functions is to use a convenient mnemonic. Here is one.

$$\sin\theta = \frac{\text{opposite}}{\text{hypotenuse}} \qquad \frac{\text{Oscar}}{\text{Had}}$$

$$\cos\theta = \frac{\text{adjacent}}{\text{hypotenuse}} \qquad \frac{\text{A}}{\text{Hand}}$$

$$\tan\theta = \frac{\text{opposite}}{\text{adjacent}} \qquad \frac{\text{Out}}{\text{Again}}$$

## STUDENT QUIZ

Name ______________________________________________

1. The sum of the three angles in a right triangle is 90°.
   a. true
   b. false
2. The longest side of a right triangle is called the hypotenuse.
   a. true
   b. false
3. The Pythagorean theorem is useful for finding the angles in a right triangle.
   a. true
   b. false
4. Trig functions are simply the ratios of various sides.
   a. true
   b. false
5. The sides of a right triangle are often called R, X, and Z in electronic circuits.
   a. true
   b. false

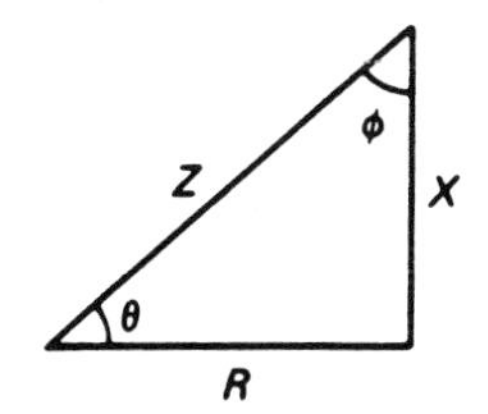


**Figure 17-1**

6. See figure 17-1. If $\theta$ =42° find the angle $\Phi$.
   a. 138°
   b. 90°
   c. 48°
   d. 42°
7. See figure 17-1. Which side is adjacent to angle $\Phi$?
   a. Z
   b. R
   c. X
8. See figure 17-1. If R = 22 kΩ and X = 90 kΩ, find Z.
   a. 92.6 kΩ
   b. 2.65 kΩ
   c. 24.6 kΩ
   d. not enough data to calculate

9. See figure 17-1. If Z = 100 kΩ and R = 47 kΩ, find X.
   a. 147 kΩ
   b. 88.3 kΩ
   c. 110.5 kΩ
   d. 53 kΩ
10. See figure 17-1. If Z = 57.4 Ω and X = 12 Ω, find R.
   a. 69.4 Ω
   b. 45.4 Ω
   c. 58.6 Ω
   d. 56.1 Ω
11. See figure 17-1. If X = 12 Ω and R = 12 Ω, find the angle $\theta$.
   a. 45°
   b. 90°
   c. 0°
   d. 22.5°
12. See figure 17-1. If $\theta$ = 33.5° and R = 4.7 kΩ, find Z.
   a. 0.55 kΩ
   b. 0.66 kΩ
   c. 5.64 kΩ
   d. 6.61 kΩ
13. See figure 17-1. If $\theta$ = 67° and Z = 88 kΩ, find R.
   a. 34.4 kΩ
   b. 0.39 kΩ
   c. 2.35 kΩ
   d. 53.6 kΩ
14. Determine tan 17.25°.
   a. 0.297
   b. 0.955
   c. 0.311
   d. 3.221
15. Find $\sin^{-1}$ 0.9814.
   a. 0.999°
   b. 0.017°
   c. 79.8°
   d. 78.9°

# TRIGONOMETRIC IDENTITIES

CHAPTER 18

## LAW OF SINES

- The ***Law of Sines*** enables us to calculate sides or angles of oblique triangles.
- If we know two sides and the included angle or if we know two angles and the side opposite one of them, the Law of Sines will let us solve for any other side or angle.
- There are three general equations in this law.

  $\sin C/c = \sin B/b$

  $\sin C/c = \sin A/a$

  $\sin A/a = \sin B/b$

- Each of these equations can be used depending on the unknown and known values.

## LAW OF COSINES

- The ***Law of Cosines*** will let us solve for sides or angles of triangles that the Law of Sines prevents, including finding the angles when only the sides are known.
- The three equations in this law are as follows.

  $a^2 = b^2 + c^2 - 2bc \cos A$

  $b^2 = a^2 + c^2 - 2ac \cos B$

  $c^2 = a^2 + b^2 - 2ab \cos C$

- Arranging either of these equations will enable you to solve for unknowns in a triangle.

## PRACTICAL COMMENTS

- The need to solve these oblique triangles arise in your study of phase relationships in certain electronic and power circuits. Work with many types of filter circuits involves using these laws to properly design the circuits frequency response.

## STUDENT QUIZ

Name ______________________________

1. The easiest law to apply to solve a right triangle if you only need to know the sides is the Pythagorean Law.
   a. true
   b. false
2. If you know two angles and the included side, then you should apply the Law of Cosines.
   a. true
   b. false
3. The Law of Sines is valuable when you need to know an angle but you only have the three sides given.
   a. true
   b. false
4. The sum of the angles in any triangle is 180°.
   a. true
   b. false
5. An oblique triangle can have one angle larger than 90°.
   a. true
   b. false

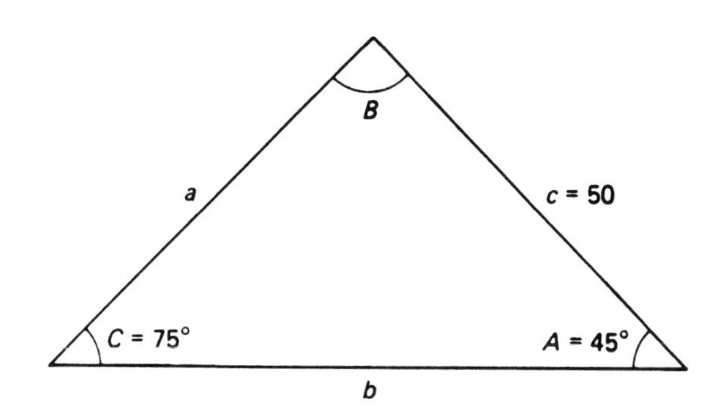


**Figure 18-1**

6. See figure 18-1. Find side b.
   a. 70.7
   b. 14.14
   c. 36.6
   d. 107.3
7. See figure 18-1. Find side a.
   a. 36.6
   b. 29.89
   c. 14.14
   d. 70.7

8. Given a triangle with a = 40, b = 50, and c = 80. Find the angle A.
   a. 155.85°
   b. 65.85°
   c. 24.15°
   d. 22.71°
9. Given a triangle with a = 40, b = 50, and c = 80. Find the angle B.
   a. 24.15°
   b. 22.71°
   c. 125.1°
   d. 30.75°
10. Given a triangle with a = 40, b = 50, and c = 80. Find the angle C.
   a. 24.15°
   b. 22.71°
   c. 125.1°
   d. 30.75°

# AC FUNDAMENTALS

CHAPTER 19

## GENERATING ANGLES

- Our previous study of rectangular coordinates gave us a method of plotting any point.
- We can plot a line in the horizontal direction on the x-axis. This line is a base line, and it has an angle of 0°.
- The vertical y-axis represents an angle of 90°.
- The negative x-axis has an angle of 180° and the negative y-axis an angle of 270°.
- It is customary to consider a line as rotating in a positive counterclockwise direction. A clockwise rotation will produce a negative angle.
- The line that is rotated is called a ***vector***, and can represent a voltage or current in an AC circuit.
- The angle between the base and our graph line is called ***theta ($\theta$)***.
- This angle is measured in degrees or *radians*. A radian = 57.3°.
- To convert an angle in degrees to radians, divide the angle by 57.3.
- The opposite conversion, from radians to degrees, just multiply the angle in radians by 57.3°.

## GENERATING SINE WAVES

- If we rotate a vector through 360°, the end of the vector will produce a sine wave.
- The positive peak value will be at 90°. The negative peak value will occur at 270°.
- The value will be zero at 0° and 180° rotation.
- The value of the output voltage can be found at any point by $v = V_{pk}\sin\theta$.

## FREQUENCY AND TIME

- The length of time used to generate one complete sine wave is called the*Period* (T). Lo case t is used to indicate any time.

- ***Frequency*** (f) is the number of complete cycles of sine waves generated in a specific am of time.

- ***Period*** and ***frequency*** are reciprocals of each other. That is,
  $f = 1 / T$ and $T = 1 / f$

## ANGULAR VELOCITY

- If the sine wave is generated at a constant rate of rotation, that rate of rotation is ca *angular velocity* ( $\omega$ ).
  $\omega = 360°f$ if you want the answer to be in degrees.
  $\omega = 2 \pi f$ if you want the answer to be in radians.
  $\omega = 400_g f$ if you want the answer to be in gons or grads.

## RMS AND PEAK VALUES OF VOLTAGE AND CURRENT

- The peak of maximum value of the sine wave is at 90°
  The RMS or effective value is at 45° and equals peak value / $\sqrt{2}$
  $I_{RMS} = I_{eff} = I_{pk} \sin 45° = I_{pk} / \sqrt{2} = .707\ I_{pk}$
  If $V_{pk} = 8\ V_{pk}$, then $V_{RMS} = 8\ V_{pk}/\sqrt{2} = 5.657\ V_{RMS} = 5.657\ V_{eff} = 5.657\ V$

- A value given without a subscript is assumed to be RMS. $9\ V = 9\ V_{RMS}$

- The peak or max value equals RMS value x $\sqrt{2}$
  $5.6\ mA = 5.6\ mA_{RMS} = 5.6\ mA_{eff} = 7.92\ mA_{pk} = 7.92\ mA_{max} = 15.84\ mA_{pp}$

## PHASE ANGLE OF LEAD OR LAG

- In an AC circuit with just a resistance, current and voltage are in phase which means 0°

- $| \Theta |$ means absolute value of $\Theta$ which means the result is always positive. $| -42° |$ = 42° and $| +42° | = 42°$

- An AC circuit which includes a capacitor and/or inductor has voltage and current ou phase with each other by phase shift of $| \Theta |$ .

- The voltage leads the current in any inductive circuit by $| \Theta |$ .
  The current leads the voltage in any capacitive circuit by $| \Theta |$ .

- In a series circuit, the current is constant. The voltage will then either lead or lag or b phase with the current.

- In a parallel circuit, the voltage is constant. The current will then either lead or lag or be in phase with the voltage.

## INSTANTANEOUS VALUES OF VOLTAGE AND CURRENT

- Lower case letters are used for instantaneous values of voltage and current and time. Be sure to include a subscript of t after the unit to show instantaneous ( $4.56\ V_t$ or $7.89\ mA_t$ ) or subscripts of RMS, eff, pk, max, pp.

- The formula for instantaneous current is: $i = I^{pk} \sin(\omega t \pm |\Theta|)$
  and for instantaneous voltage is: $v = V^{pk} \sin(\omega t \pm |\Theta|)$

- Assume a parallel inductive circuit with given values of
  V = 12 V, I = 5 mA, frequency = 450 Hz, t = 2 m sec, $\Theta$ = -40°
  voltage is the reference so $\Theta = 0°$ in voltage equation.
  voltage leads current by 40° so $\Theta = -40°$ in current equation.
  The general formulas for instantaneous current and voltage are:
  $i = 5 \times \sqrt{2}\ mA^{pk} \sin(\omega t - 40°) = 7.071\ mA^{pk} \sin(\omega t - 40°)$
  $v = 12 \times \sqrt{2}\ V^{pk} \sin(\omega t + 0°) = 16.97\ V^{pk} \sin(\omega t)$
  Notice that $\omega t$ is left as $\omega t$ in general equations.
  $\omega = 360°f = 2\pi f = 400_g f$
  Since $\Theta$ is given in degrees, calculate $\omega t$ in degrees.
  $\omega t = \frac{360°}{\text{cycle}} \times \frac{450\ \text{cycles}}{\text{sec}} \times 2\ \text{msec} = 324°$
  $i = 7.071\ mA^{pk} \sin(324° - 40°) = -6.861\ mA_t$
  $v = 16.97\ V^{pk} \sin(324°) = -9.975\ V^t$

# PRACTICAL COMMENTS

- The oscilloscope is nothing more than a graph showing the instantaneous voltage in a circuit. Now you can begin to see how the scope can be useful. While the scope does not measure current directly, it can give an indication of the current in an AC circuit. Properly connected, the scope can show the instantaneous voltage and current in any AC circuit. If the circuit has only resistance, then the traces on a dual trace scope will be in phase. Using a capacitor or inductor will enable you to see and measure the phase difference between the voltage and the current.

- Be sure to use proper subscripts (RMS, eff, pk, max, pp, t) on units so there is no question which is which.

## STUDENT QUIZ

Name ______________________________________________

1. A rotating constant vector quantity will produce a sine wave.
   a. true
   b. false
2. There are 5.35°/radian.
   a. true
   b. false
3. The instantaneous value of voltage in an AC circuit can be calculated by $v = 2\pi f$.
   a. true
   b. false
4. The voltage and current in an AC circuit that includes a capacitor are out of phase.
   a. true
   b. false
5. The statement $V_{pk} = 0.707\ V_{RMS}$ is
   a. true
   b. false
6. If the period equals 15 $\mu$s, find f.
   a. 15 kHz
   b. 258 Hz
   c. 66.6 kHz
   d. 66.6 Hz
7. Convert 28° to radians.
   a. 1604 radians
   b. 0.489 radians
   c. 60 radians
   d. 0.017 radians
8. Convert 2.75 radians to degrees.
   a. 11145°
   b. 0.412°
   c. 1705°
   d. 158°
9. Find the angular velocity in degrees per second for a frequency of 37.5 Hz.
   a. 13500°/s
   b. 235°/s
   c. 15000°/s
   d. 9.6°/s
10. Find the frequency if the angular velocity is 4000°/s.
    a. 0.09 Hz
    b. 11.1 Hz
    c. 111 Hz
    d. 1.11 Hz

11. If V = 20 $V_{pk}$, and f = 2 kHz, find the instantaneous voltage, v after 25 $\mu$s.
    a. 0.309 V
    b. 3.09 V
    c. 6.18 V
    d. 13.8 V
12. An AC voltage is 18 $V_{RMS}$. Convert this to $V_{pk}$.
    a. 50.92 $V_{pk}$
    b. 25.5 $V_{pk}$
    c. 17.73 $V_{pk}$
    d. 5.56 $V_{pk}$
13. An AC voltage is 369 $V_{p\text{-}p}$. Convert this to RMS.
    a. 284 $V_{RMS}$
    b. 521 $V_{RMS}$
    c. 65 $V_{RMS}$
    d. 130 $V_{RMS}$
14. An AC current is 700 mA. Convert this to $I_{pk}$.
    a. 990 mA
    b. 495 mA
    c. 248 mA
    d. 124 mA
15. An AC voltage is 125 $V_{pk}$. Convert this to $V_{p\text{-}p}$.
    a. 177 V
    b. 88.4 V
    c. 353 V
    d. 250 V

# AC CIRCUIT ANALYSIS: SERIES CIRCUIT

CHAPTER 20

## SERIES RC CIRCUITS

- Every RC circuit has some resistance and some capacitance.
- In a series circuit the current is the same in all parts of the circuit.
- The voltage across the resistor and current through it are in phase with each other.
- The voltage across the capacitor and the current through the capacitor are not in phase. The capacitor current leads the voltage by 90°.
- Drawing a vector diagram of the current in the circuit, we draw its vector horizontally. This vector represents I and $V_R$.
- The vector for $V_C$ is drawn down at 90°.
- The resultant is the source voltage V. The ***phase angle*** ***($\theta$)*** is the angle between the source V and the current I.
- The phase angle can be calculated using trig functions.
- Since there is only one current in the circuit, the same vectors can represent the resistance, reactance, and impedance.
- The capacitor offers an opposition to the flow of AC current. This is called ***capacitive reactance*** ***($X_c$)***.
- Capacitive reactance is found by $X_C = \frac{1}{2\pi fC}$. The unit is ohms.

## SERIES RL CIRCUITS

- In a circuit with an inductor and resistor in series, the inductance causes the voltage to lead the current.
- We plot the current and $V_R$ along the x-axis. The voltage across the inductor leads the current by 90°.

- The resultant of these two vectors is the source voltage.
- The source voltage leads the current by the phase angle $\theta$.
- The resulting right triangle can be solved using the Pythagorean theorem and trig functions.
- The opposition to the flow of current in an inductor is called ***inductive reactance ($X_L$)***. The unit is ohms.
- The formula for finding inductive reactance is $X_L = 2\pi fL$.
- There is an important ratio used with inductance. This ratio is called ***Q***. The ratio is shown as follows: $Q = X_L/R$.

**POLAR TO RECTANGULAR CONVERSION**

- The vector for $V_S$ is expressed with a magnitude and the phase angle. In polar form the vector might be $45\angle{-30^\circ}$ In rectangular form it might be 34.9 - j25.
- A calculator is used to convert from polar to rectangular and vice versa.
- Vectors which lie along the same plane can be added or subtracted directly.
- To add vectors which are not in the same plane, convert the vectors to rectangular form then add these coordinates and convert back to polar form.
- Subtraction follows a similar pattern.
- Series RLC circuits can be combined using addition of the rectangular coordinates Remember the L and C vectors are in the same plane but opposite in polarity.

**SERIES RESONANCE**

- In a series RLC circuit there is a frequency where $X_L = X_C$. This condition is called ***resonance***.
- At resonance, $\theta = 0°$ and $Z = R$. The impedance is minimum and the current is maximum.
- The ***resonant frequency ($f_r$)*** is found by $f_r = \dfrac{1}{2\pi\sqrt{LC}}$.
- The ratio Q is useful in determining $V_C$ and $V_L$ as follows: $Q = V_c/V_S$ or $V_C = QV_S$. Also $V_L = QV_S$.

## PRACTICAL COMMENTS

- The use of vector diagrams to represent the voltages or currents in an AC circuit saves time and will give you a good understanding of AC circuit theory. The phase relationships between the voltage and current in series RLC circuits are used extensively in the design and maintenance of such circuits as filters, transmitters, television, and medical equipment. Communication circuits of all kinds use the principle of resonance to operate.

## STUDENT QUIZ

Name ______________________________________________

1. The phase angle in a series circuit is the angle between the resistance and the source voltage.
   a. true
   b. false
2. Vectors can be expressed in polar or rectangular form.
   a. true
   b. false
3. At resonance, a series RLC circuit has $X_L = X_C$.
   a. true
   b. false
4. The vector sum of the voltage drops around a series RC circuit is the source voltage.
   a. true
   b. false
5. Any series AC circuit can be simplified into a circuit containing a resistance, an inductance, or capacitance.
   a. true
   b. false

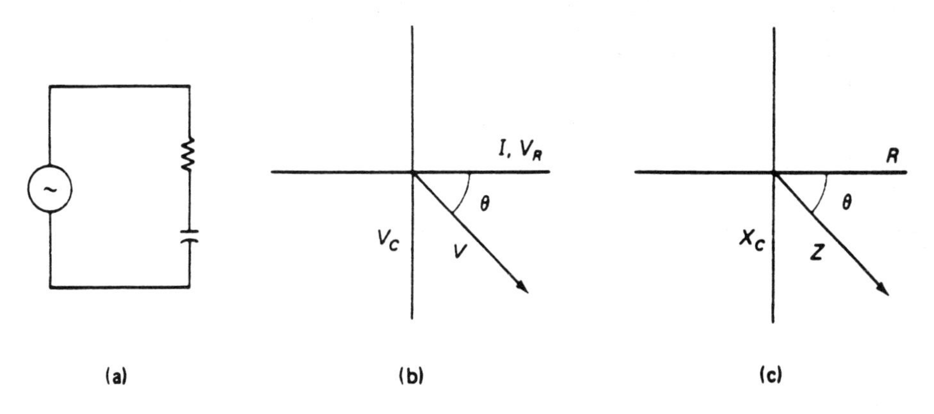


**Figure 20-1**

6. See figure 20-1. If $V_R = 18$ V and $V_C = 12$ V, find $V_T$ and $\theta$.
   a. 21.63 V, 56.3°
   b. 36 V, 33.7°
   c. 21.63 V, 33.7°
   d. 56.3 V, 33.7°

7. See figure 20-1. If V = 60 V and $\theta$ = 22°, find $V_R$ and $V_C$.
   a. 30 V, 45 V
   b. 58.9 V, 32 V
   c. 12.3 V, 5.68 V
   d. 22.5 V, 55.6 V
8. See figure 20-1. If R = 47 kΩ, C = 0.022 μF, and f = 300 Hz, find Z.
   a. 52.8 kΩ
   b. 5.8 kΩ
   c. 5.28 kΩ
   d. 5.28 MΩ
9. See figure 20-1. If R = 68 kΩ, C = 2500 pF, f = 22 kHz, and V = 18 V, find I.
   a. 26.4 mA
   b. 68.1 mA
   c. 0.264 mA
   d. 28.95 mA

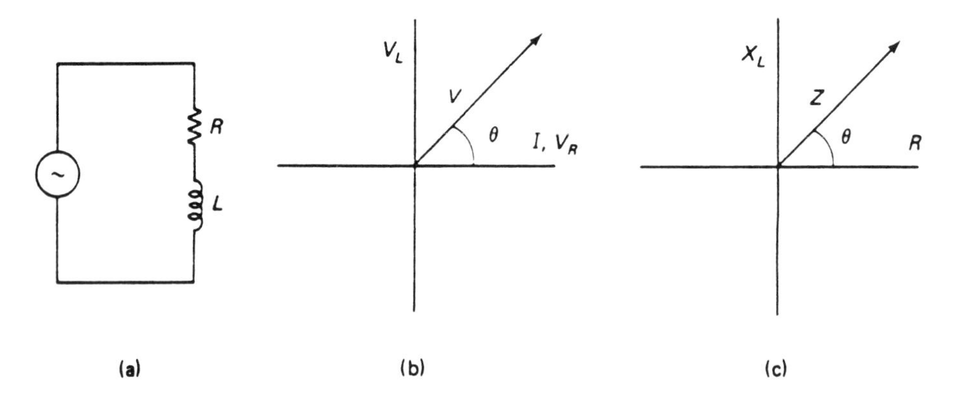


**Figure 20-2**

10. See figure 20-2. If $V_R$ = 18 V and $V_L$ = 20 V, find V.
    a. 26.9 V
    b. 38 V
    c. 12 V
    c. 12.1 V
11. See figure 20-2. If $\theta$ = 26° and V = 35 V, find $V_L$.
    a. 31.45 V
    b. 15.34 V
    c. 44.79 V
    d. 16.11 V
12. See figure 20-2. If $X_L$ = 680 Ω and $\theta$ = 27°, find Z.
    a. 1335 Ω
    b. 680 Ω
    c. 1498 Ω
    d. 655 Ω

13. What is the result of adding the following impedances:
$Z_1 = 4.7\angle 25^\circ$ kΩ, $Z_2 = 6.8\angle 32^\circ$ kΩ.
   a. $10.01\angle 32^\circ$ kΩ
   b. $11.47\angle 5^\circ$ kΩ
   c. $5.59\angle 10.23^\circ$ kΩ
   d. $11.47\angle 11.47^\circ$ kΩ

14. Express 5.76 - j9.2 in polar form.
   a. $10.9\angle 45.7^\circ$
   b. $10.9\angle 57.9^\circ$
   c. $10.9\angle -57.9^\circ$
   d. $57.9\angle -10.9^\circ$

15. Express $9.76\angle 34^\circ$ in rectangular form.
   a. 8.09 - j5.46
   b. 8.08 + j5.46
   c. 5.46 - j8.09
   d. 5.46 + j8.09

# AC CIRCUIT ANALYSIS: PARALLEL CIRCUITS

CHAPTER 21

## RC CIRCUIT ANALYSIS

- In a parallel circuit the voltage across each branch is equal.
- The voltage across and the current through the resistor are plotted along the horizontal axis.
- The current through the capacitor is plotted along the positive y-axis.
- The total current is the vector sum of $I_L$ and $I_R$.
- $X_C$, Z, and R are not ususally plotted for a parallel circuit; however their reciprocals are often plotted. The reciprocal of $X_C$ is ***susceptance (B)***. The reciprocal of impedance is ***admittance (Y)***. The reciprocal of resistance is ***conductance (G)***.

## RL CIRCUIT ANALYSIS

- As in an RC circuit, the source voltage and resistor current, in an RL circuit, are plotted horizontally.
- The current through the inductor is plotted down at -90°.

## EQUIVALENT CIRCUITS

- To multiply or divide vectors they must be in polar form.
- Multiply or divide the magnitudes. Add the angles if you are multiplying; subtract the angles if you are dividing.
- A parallel circuit can be converted to an equivalent series circuit.
- The admittance of the parallel circuit is the reciprocal of the impedance of the series circuit. Make the conversion. Find R and $X_C$.

- In many complex series-parallel circuits it is necessary to make many conversions between series and parallel.

- A group of formulas is often used which are called ***transforms***.

- The transforms from series to parallel are $G = \frac{R}{R^2 + X^2}$ and $B = \frac{X}{R^2 + X^2}$.

- To convert from parallel to series, the transforms become $R = \frac{G}{G^2 + B^2}$ and $X = \frac{B}{G^2 + B^2}$.

- These transform equations give absolute values. The correct signs must be added.

## PRACTICAL COMMENTS

- Currents can be added in parallel circuits because the voltage is a constant. It is easy to calculate for the total current in a parallel circuit and then use that value to solve for the circuit's impedance using Ohm's law. This method is often less prone to errors that to use B and Y in a parallel circuit. Either method will result in the correct answer. Use the method that will be simpler and less work for you. The easier method will result in more correct answers.

## STUDENT QUIZ

Name ______________________________________________

1. In a parallel RC circuit the vector for the voltage and the $I_T$ is drawn in the same plane.
   a. true
   b. false
2. A parallel RL circuit has the voltage and the current through the resistor in phase.
   a. true
   b. false
3. To divide vectors they must be in rectangular form.
   a. true
   b. false
4. To multiply vectors, the angles are added.
   a. true
   b. false
5. Transform formulas are used to convert from series to parallel equivalent circuits.
   a. true
   b. false

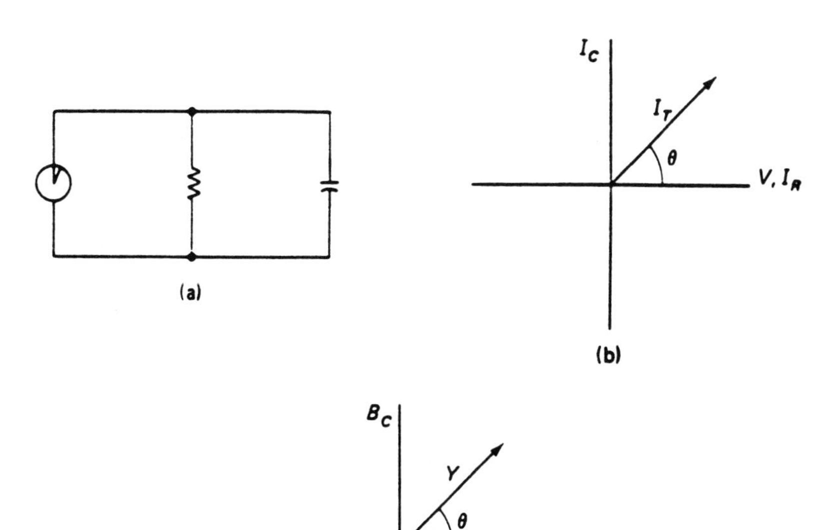


**Figure 21-1**

6. See figure 21-1. If $I_T$ = 4.7 mA and $\theta$ = 42.5°, find $I_C$.
   a. 3.47 mA
   b. 1.52 mA
   c. 1.23 mA
   d. 3.18 mA
7. See figure 21-1. If $I_R$ = 22.57 mA and $I_C$ = 32 mA, find $\theta$.
   a. 35.2°
   b. 44.9°
   c. 54.8°
   d. 74.9°
8. See figure 21-1. If $I_R$ = 23 mA, $I_C$ = 12.8 mA, and $V_T$ = 18 V, find Z.
   a. 1.46 kΩ
   b. 0.68 kΩ
   c. 780 Ω
   d. 1.41 kΩ

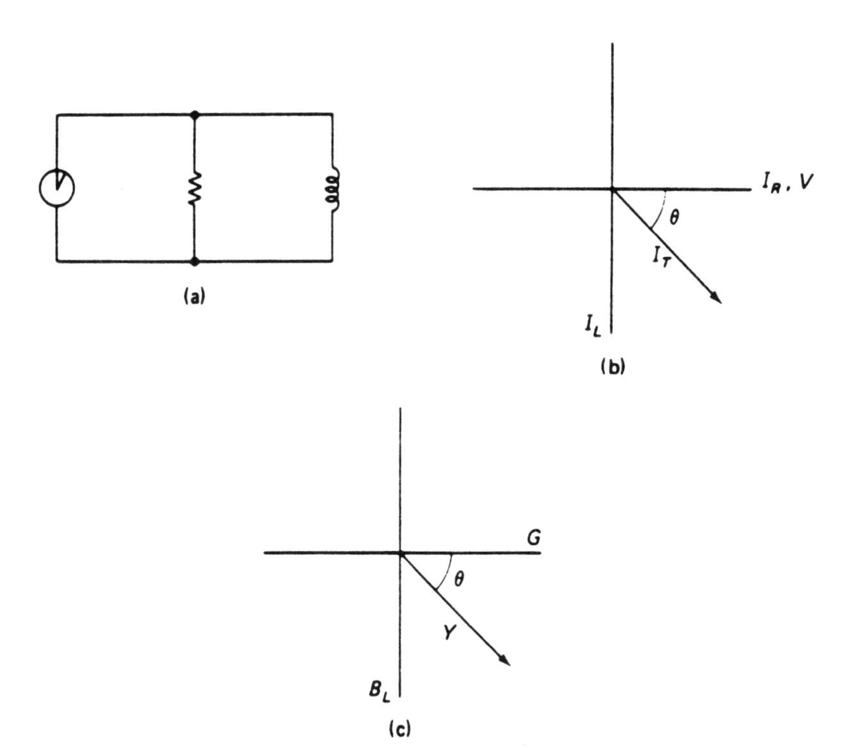


**Figure 21-2**

9. See figure 21-2. If $V_T$ = 12 V, $I_T$ = 75 mA, and $\theta$ = -19°, find Z.
   a. 160 Ω
   b. 737 Ω
   c. 253 Ω
   d. 229 Ω

10. See figure 21-2. If $G = 492\ \mu S$ and $B = 250\ \mu S$, find the resistance R of the equivalent series circuit.
    a. 2033 Ω
    b. 4000 Ω
    c. 1615 Ω
    d. 1234 Ω

# FILTERS

CHAPTER 22

## LOW-PASS AND HIGH-PASS FILTERS

- A ***low-pass filter*** is a circuit that passes low frequencies easily and attenuates higher frequencies.

- To attenuate means to reduce the amplitude or peak-to-peak value of the voltage.

- A low-pass filter can consist of a resistor in series with a capacitor. The output $V_O$ is taken across the capacitor.

- There is a frequency where $X_C = R$. This is called the ***cutoff frequency*** **($f_{co}$)**.

- The cutoff frequency is found by $f_{co} = 1/2\pi RC$. This formula is the same for either a high-pass or a low-pass filter.

- $V_O$ at the cutoff frequency equals $0.707V_{in}$. $V_O = 0.707V_{in}$.

- The phase angle $\theta$ is -45° at $f_{co}$.

- The power is $P = 0.5\ P_{max}$.

- Reversing the positions of the resistor and capacitor will result in a high-pass filter.

- A ***high-pass filter*** passes high frequencies easily and attenuates low frequencies.

- All of the above formulas and conditions are also true for an RC high-pass filter.

## THÉVENIN'S THEOREM

- Thévenin's theorem can be used to solve some AC filter circuits.

- A low-pass filter with several resistors may require this theorem to actually find the equivalent resistance. You can then proceed to find the necessary values.

- Norton's theorem can also be used in solving filter circuits.

## BAND-PASS FILTERS

- A ***band-pass filter*** passes a band of frequencies easily but rejects all frequencies above and below the ***pass-band.***
- The pass-band is that group of frequencies between the upper and lower cut off frequencies.
- A band-pass filter can consist of a high-pass filter circuit followed by a low-pass circuit. Putting the filters opposite positions will also produce a band-pass result.
- Thévenin's and Norton's theorems can be used to find values at the cut off frequencies.
- A band-pass filter can also be made using an RLC circuit.
- An inductor and capacitor connected in series would form a resonant circuit. The output would be taken across a series resistance.
- The ***bandwidth (BW)*** is the same as the pass-band. It is found as follows, $BW = f_2 - f_1$.
- The resonant frequency falls essentially in the middle of the band so $f_1 = f_r - BW/2$ and $f_2 = f_r + BW/2$.
- There is also a relationship between $f_r$, BW, and Q. It is $BW = f_r/Q$.
- A ***band-stop*** filter is a filter that will reject frequencies in a band and pass the frequencies above and below the cutoff frequencies.
- Another name for a band-stop filter is ***band-reject.***
- A band-reject filter usually has a parallel resonant circuit in series with a resistance. $V_o$ is taken across the resistance.

## PRACTICAL COMMENTS

- A band-pass RC filter can have either section first in the signal path; however it is good design procedure to have the low-pass section first. Loading effects are minimized with this configuration. RC filters are usually used at audio frequencies. Filters that operate at radio frequencies are usually of the LC type.

## STUDENT QUIZ

Name ______________________________________________

1. A filter that passes low frequencies easily is called a low-pass filter.
   a. true
   b. false
2. A high-pass filter attenuates high frequencies.
   a. true
   b. false
3. The BW of a band-pass filter is found by $f_2 - f_1$.
   a. true
   b. false
4. A band-reject filter passes all frequencies between the two cutoff frequencies.
   a. true
   b. false
5. At the cutoff frequency of a low-pass filter $X_C = R$.
   a. true
   b. false

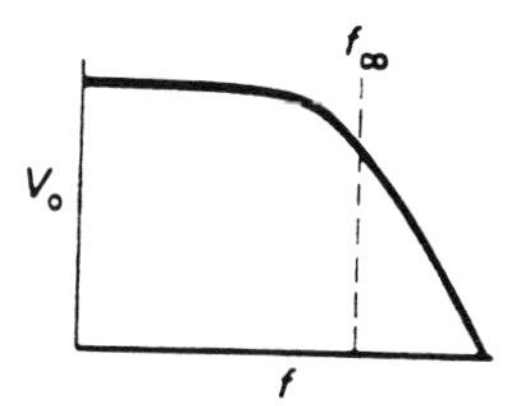


**Figure 22-1**

6. See figure 22-1. This represents the frequency response curve of the output from a
   a. high-pass filter.
   b. low-pass filter.
   c. band-reject filter.
   d. band-pass filter.

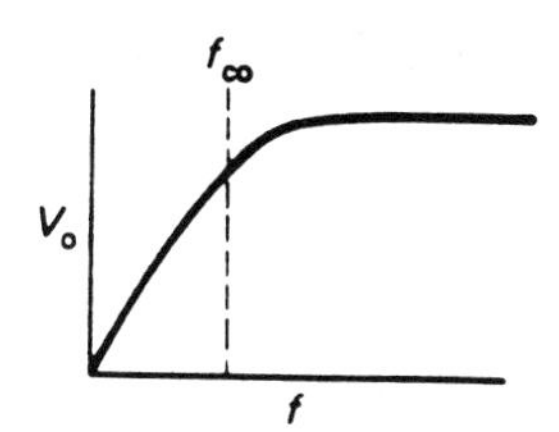


**Figure 22-2**

7. See figure 22-2. This represents the frequency response curve of the output from a
   a. high-pass filter.
   b. low-pass filter.
   c. band-reject filter.
   d. band-pass filter.

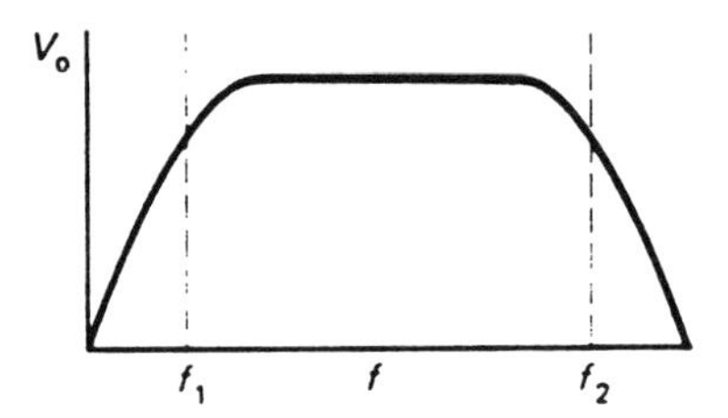


**Figure 22-3**

8. See figure 22-3. This represents the frequency response curve of the output from a
   a. high-pass filter.
   b. low-pass filter.
   c. band-reject filter.
   d. band-pass filter.

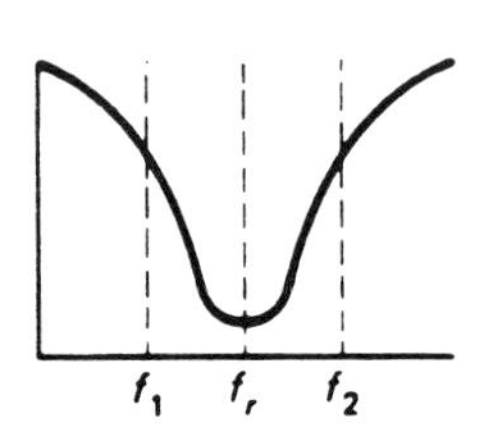


**Figure 22-4**

9. See figure 22-4. This represents the frequency response curve of the output from a
   a. high-pass filter.
   b. low-pass filter.
   c. band-reject filter.
   d. band-pass filter.

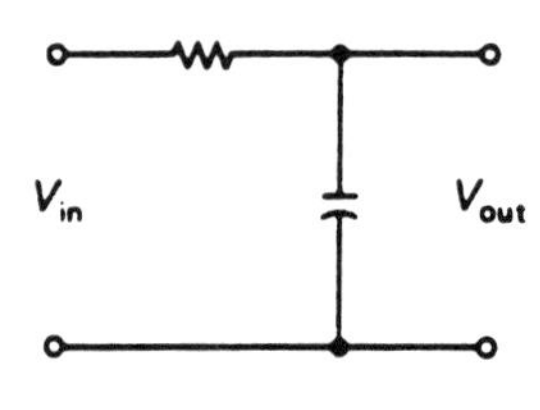


**Figure 22-5**

10. See figure 22-5. If R = 47 kΩ and C = 256 pF, find $f_{co}$.
    a. 13.2 kHz
    b. 132 kHz
    c. 1320 Hz
    d. 132 Hz
11. See figure 22-5. If R = 520 kΩ, find the value of C that will give an $f_{co}$ of 100 kHz.
    a. 3.06 μF
    b. 1.59 μF
    c. 3.06 pF
    d. 1.59 pF

12. See figure 22-5. If R = 10 kΩ, C = 22 μF, and $V_{in}$ = 17 V, find $V_{out}$ at $f_c$.
    a. 12 V
    b. 17 V
    c. 9.25 V
    d. 7.12 V

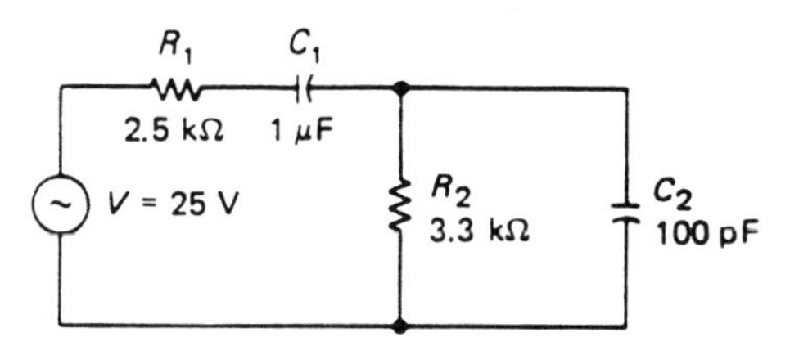


**Figure 22-6**

13. See figure 22-6. This circuit is known as a
    a. low-pass filter,
    b. high-pass filter.
    c. band-pass filter.
    d. band-reject filter.
14. See figure 22-6. Find $f_1$.
    a. 1.12 MHz
    b. 27.5 Hz
    c. 5.55 kHz
    d. 74.4 Hz
15. See figure 22-6. Find $f_2$.
    a. 1.12 MHz
    b. 27.5 Hz
    c. 5.55 kHz
    d. 74.4 Hz

# LOGARITHMS

CHAPTER 23

## LOGARITHMS DEFINED

- Remember that $10_3$ = 1000. The base is 10 and the exponent is 3. The ***logarithm*** of the expression is the exponent 3.

- A logarithm is an exponent to which a base must be raised to equal a number.

## COMMON LOGARITHMS

- ***Common logarithms*** use the base of 10.

- A common logarithm is written log 1000 = 3.

- The expression $10_3$ = 1000 is an ***exponential expression.***

- A ***logarithmic expression*** would be log 1000 = 3.

- Common logarithms of numbers between 1 and 10 can be found in a table of common logarithms. Your scientific calculator can also present these values.

- Common logarithms of any number can easily be found on your calculator. Try this one: what is the log of 75? Your calculator should show that the log 75 = 1.875.

- If the number is a decimal, the logarthim is a negative exponent. For example, log 0.075 = -2.125.

## MULTIPLICATION OF NUMBERS BY USING LOGARITHMS

- Since a common logarithm is an exponent of the common base 10, the exponents are added to multiply.

- An example is as follows:
  102 X 75 = log 102 + log 75 = log 3.884 = 7650

## DIVISION OF NUMBERS BY USING LOGARITHMS

- Division of numbers using logarithms involves subtracting the exponents Here is an example: 102/75 = log102 - log75 = 1.36

## RAISING A NUMBER TO A POWER BY USING LOGARITHMS

- This process involves multiplying the exponents. An example of this is $6^3 = 3 \log 6 = 216$.

## NATURAL LOGARITHMS

- A ***natural logarithm*** is an exponent where the base is (e).

- The numerical value of e is rounded to 2.72.

- Natural logarithms can easily be found using your calculator.

- The abbreviation for natural logarithms is ln.

# PRACTICAL COMMENTS

- Logarithms are a convenient method of doing calculations by hand. With the advent of the calculator, the hand method has gone out of style. Logs, as they are abbreviated, are used extensively in electronic work, as you will see in the next chapter.

**STUDENT QUIZ**

Name ______________________________________________

1. This expression log 0.0089 = -2.05 is a logarithmic expression.
   a. true
   b. false
2. Logarithms are exponents.
   a. true
   b. false
3. The addition of logs will produce a division of the numbers.
   a. true
   b. false
4. If the log is negative, the number must be greater than one.
   a. true
   b. false
5. Natural logarithms are to the base e.
   a. true
   b. false
6. Find the common logarithm of 32.765.
   a. 3.489
   b. $5.82 \times 10^{32}$
   c. 1.515
   d. 0.644
7. Perform the following using logarithms: 28.9 X 54.6
   a. 345.03
   b. 584.36
   c. 0.529
   d. 1577.94
8. Perform the following using logarithms: 1234/19.56
   a. 24137
   b. 63.088
   c. 41.44
   d. 1.799
9. Perform the following using logarithms: $0.00345^{-1.9}$
   a. 47659
   b. 19.764
   c. -19.764
   d. 4.678
10. Find the number whose natural logarithm is 0.693.
    a. -0.159
    b. 9.18
    c. 2
    d. 1.765

# LOGARITHMIC EQUATIONS

CHAPTER 24

## LOGARITHMIC AND EXPONENTIAL FORMS

- The two forms of equations used are the exponential and the logarithmic.
- The basic logarithmic equation is $\log N = x$ or $\ln a = b$.
- An exponential equivalent would be $10^x = N$ and $e^b = a$.

## LOGARITHMIC EQUATIONS: COMMON LOGS

- An example of solving a logarithmic equation is $x = \log 750$. Finding $\log 750 = 2.875$, we know that $x = 2.875$.
- The exponential form of the same is $\log a = 2.875$. The solution is $7.5 \times 10^2$.
- An equation such as, $x = 9.3^{1.2}$ can be solved by taking the log of both sides of the equation as follows, $\log x = 1.2 \log 9.3$, from which we find that $x = 14.53$.

## LOGARITHMIC EQUATIONS: NATURAL LOGS

- Logarithmic equations can use natural logs. An example is: $e^{2.1} = x$. This can be solved by taking the inverse natural log of 2.1 which = 8.166.
- The calculator is invaluable in these types of problems.

## PRACTICAL COMMENTS

- Many types of electronic circuits operate on a logarithmic response. An example is the use of decibels, a unit to measure power level changes. Another type of circuit is the curve that results when you charge a capacitor. The increase in voltage across the capacitor follows a natural log curve.

## STUDENT QUIZ

Name ______________________________

1. An equation in logarithmic form is written $x = \log N$.
   a. true
   b. false
2. An equation written in exponential form is $x = a^2 + b$.
   a. true
   b. false
3. Common logarithms have a base of 10.
   a. true
   b. false
4. Natural logarithms have a base of x.
   a. true
   b. false
5. A logarithmic equation must be solved using the same rules as any equation.
   a. true
   b. false
6. Solve the following equation: $\log x = 0.375$
   a. $x = 7.32$
   b. $x = -0.981$
   c. $x = 0.019$
   d. $x = 2.37$
7. Solve the following equation: $\log 0.06753 = y$
   a. $y = -2.695$
   b. $y = -1.171$
   c. $y = 1.168$
   d. $y = 0.155$
8. Solve the following equation: $x = 10^{1.2}$
   a. $x = 15.85$
   b. $x = -15.85$
   c. $x = 3.32$
   d. $x = -3.32$
9. Solve the following equation: $5 = \log (x/75)$
   a. $x = 1.33 \times 10^{-7}$
   b. $x = 7.5 \times 10^{6}$
   c. $x = 6.875$
   d. $x = -7.5 \times 10^{-6}$
10. Solve the following equation: $\ln y = 1.50$
    b. $y = 31.6$
    b. $y = -31.6$
    c. $y = 4.48$
    d. $y = 2.72$

# APPLICATIONS OF LOGARITHMS

CHAPTER 25

## GAIN MEASUREMENTS

- An important parameter in many electronic circuits is ***gain***.
- There are several types of gain: ***voltage gain $A_V$), current gain ($A_I$),*** and ***power gain ($A_P$).***
- Gain is the ratio of output/input. It can be expressed as a ratio; for example, the voltage gain is 17, which is an expression of the ratio 17/1.
- Power and other gains are is often expressed in ***decibels (dB)***.
- The formula for power gain in dB is $A_P(dB) = 10\ log\ \frac{P_{OUT}}{P_{IN}}$
- Voltage gain is also expressed in dB: $A_V(dB) = 20log\ \frac{V_{out}}{V_{in}}$
- The formula for current gain is $A_I(dB) = 20\ log\ I\frac{out}{I_{in}}$.
- These gain formulas can be solved using logarithmic and algebraic methods previously covered.
- A typical example is the formula 20 dB = 10 log $P_{out}$/500 $\mu$W. To solve this, divide both sides of the equation by 10, change the form to exponential, and solve using algebra. $P_{out}$ = 0.05 W is the answer.
- If the dB figure is a loss (-), then the answer will indicate that $P_{out}$ is smaller than $P_{in}$.

## REFERENCE LEVELS

- A very common reference level for power gain measurement is to use a reference level of 1 mW. This reference level is called a ***dBm***.
- Using dBm's will give you actual power levels.

## FREQUENCY RESPONSE

- Most systems have a response that varies as the frequency of operation is changed.
- A ***Bode plot*** is a graph of the frequency response of a filter.
- The Bode plot shows how gain varies with frequency.
- This graph is plotted on semilog paper with frequency on the horizontal axis and gain on the vertical axis.
- The Bode plot assumes that the gain falls off or increases at a rate of ***6 dB per octave or 20 dB per decade.***
- An octave is a doubling or halving of the frequency.
- A decade changes the frequency by a factor of 10.
- Plot the points on your Bode plot starting at the cutoff frequency. Recall that $f_{co} = 1/2\pi RC$.
- The gain will drop by 20 dB for each decade of frequency change.
- Connect each two points with a straight line. The result is a Bode plot.
- There is some error in the Bode plot especially close to the cutoff frequency.

## RC CIRCUITS

- In an earlier chapter we used the universal time constant curve to determine the voltage across a capacitor as it charges.
- The actual formula for this curve is given by $v_C = \frac{V}{1 - e^{-t/RC}}$. $v_c$ is the voltage across the capacitor, V is the source voltage, t is the time in seconds, and R and C are the values of the components.
- A similar formula for the voltage across the resistance is $v_R = Ve^{-1t/RC}$.

# PRACTICAL COMMENTS

- The fact that a capacitor exhibits the effect of taking time to charge after a voltage is applied is paramount in the operation of countless electronic circuits. All of our circuits that provide communications rely on this effect. These include radio, tv, satellite communications, and many others. Telephone systems are a large user of filters for squeezing many conversations onto one pair of wires. In fact it is hard to find some electronic circuits that do not use this capacitance effect, especially since all AC-DC power supplies use capacitors to filter and smooth the pulsating AC to DC, and most electronic equipment uses power supplies of one type or another. Capacitors are in all of them.

## STUDENT QUIZ

Name ______________________________________________

1. A measure of the performance of an amplifier is often the voltage gain.
   a. true
   b. false
2. Amplifier gains must be expressed in decibels.
   a. true
   b. false
3. Power gain is the product of the voltage and current gains.
   a. true
   b. false
4. A Bode plot is a graphical representation of the frequency response of a filter.
   a. true
   b. false
5. Decibels are a convenient method of expressing the power in-out functions of a filter.
   a. true
   b. false
6. An amplifier has a $P_{in} = 3$ mW. It's $P_{out} = 13$ W. Find the power gain in dB.
   a. 72.7 dB
   b. 26.4 dB
   c. 36.4 dB
   d. 83.7 dB
7. An amplifier's $A_p = 35$ dB and its $P_{in} = 75$ mW. Find $P_{out}$.
   a. 118.5 W
   b. 237 mW
   c. 4.77 W
   d. 237 W
8. An amplifier has a power gain of 16 dBm. find $P_{out}$.
   a. 39.8 mW
   b. 79.6 mW
   c. 39.8 W
   d. 14 mW
9. An amplifier has $R_{out} = R_{in}$. $V_{in} = 500$ mV and $V_{out} = 12$ V. Find $A_v$.
   a. 55.21 dB
   b. 32.4 dB
   c. 27.6 dB
   d. 13.8 dB
10. If a filter's total response is -23 dB at the second decade frequency, find the response at the cutoff frequency.
    a. -6 dB
    b. -3 dB
    c. -20 dB
    d. -43 dB

# BOOLEAN ALGEBRA

CHAPTER 26

## LOGIC VARIABLES

- ***Boolean algebra*** is the mathematics of digital logic circuits.
- This algebra provides a system of manipulating ***logic variables***.
- Logic variables consist of states where a statement is either true or false. No other states are allowed.
- These logic states are usually called a high (logic 1) or a low (logic 0).
- In Boolean algebra we deal only with variables which are 1's or 0's.

## THE AND FUNCTION

- A basic function can be simulated by connecting two switches in series with a light and a source. You can see that the only way to turn the light on is to have both switches on.
- This type of function is called ***AND logic***.
- A table showing the various combinations of logic inputs and the outputs from an AND logic circuit is called a ***truth table***.
- Two inputs A and B are ANDed together to form an output A•B.
- We can also say that A and B = AB. Note: A AND B is a product operation.
- These logic circuits are called ***gates***. Thus we have an AND gate.
- An AND gate can have several inputs. The expression for the output is ABC if there are three inputs.

## THE OR FUNCTION

- Another basic gate is the ***OR gate***.

- This gate has the logic that if any input is a logic high, then the output will be a logic high. The only time a logic low output will occur is if all the inputs are low.

- If there are two inputs A and B and they are ORed together, the output is A + B. The mathematical operation is one of addition, but we say A is ORed with B.

- As with the AND gate, an OR gate can have several inputs.

## THE NOT FUNCTION

- A logic function that produces a ***complement*** of a single input is called a ***NOT function***.

- The logic circuit that will produce this is called an ***inverter***.

- Since the output is not equal to the input, but is the complement of it, it is called "not A" or simply $\overline{A}$.

## BOOLEAN EXPRESSIONS

- These three logic gates, AND, OR, and NOT comprise the major parts of a logic circuit.

- A logic circuit can have several variables denoting the operation desired in a logic circuit. Examples of these are BC + DE + $B\overline{A}$

and $A\overline{B}CD\overline{E}$.

- The Boolean expression can be used to draw a logic circuit to perform some desired operation.

- Keep in mind the type of mathematical operation the various gates will perform.

- A Boolean expression may also be written from a logic circuit.

## BOOLEAN POSTULATES AND THEOREMS

- A ***postulate*** is a self-evident truth. An example of this is as follows: if A = 1 then $\overline{A} = 0$.

- These postulates can be summarized as follows.

Postulates

| | |
|---|---|
| $A = 1$ (if $A \neq 0$) | $A = 0$ (if $A \neq 1$) |
| $0 \cdot 0 = 0$ | $1 \cdot 1 = 1$ |
| $1 \cdot 1 = 1$ | $1 + 1 = 1$ |
| $1 \cdot 0 = 0$ | $1 + 0 = 1$ |
| $\overline{1} = 0$ | $\overline{0} = 1$ |

**Table 26-1**

- Some algebraic properties also apply to Boolean expressions.
  The commutative property: AB = BA or A + B = B + A.
  The associative property: A(BC) = AB(C) or
  A + (B + C) = (A + B) + C.
  The distributive property: A(B + C) = AB + AC.

- There are several Boolean theorems that are used to simplify logic circuits. These theorems are summarized in Table 26-2.

Theorems

| | |
|---|---|
| $A \cdot 0 = 0$ | $A + 0 = A$ |
| $A \cdot 1 = A$ | $A + 1 = 1$ |
| $A \cdot A = A$ | $A + A = A$ |
| $A \cdot \overline{A} = 0$ | $A + \overline{A} = 1$ |
| $\overline{\overline{A}} = A$ | $A = \overline{\overline{A}}$ |

**Table 26-2**

- These theorems combined with the postulates are used to simplify a logic circuit into is simplest form. This means one must find a combination which performs the desired logic with the fewest gates.

- ***DeMorgan's theorem*** gives us a very valuable tool for simplifying logic circuits. This theorem tells us that $\overline{A}\ \overline{B} = \overline{A + B}$. This also tells us that $\overline{AB} = \overline{A} + \overline{B}$.

- DeMorgan's theorem can be simplified to break the bar and change the sign. Consider $\overline{\overline{A} + \overline{B}} = AB$ as a sample.

- The ***absorption theorem*** gives us some very common substitutions useful in simplifying Boolean expressions. These are shown in Table 26-3.

Absorption Theorem

| | |
|---|---|
| $A(A + B) = A$ | $A + AB = A$ |
| $A(\overline{A} + B) = AB$ | $A + \overline{A}B = A + B$ |

**Table 26-3**

## PRACTICAL COMMENTS

- The study of Boolean algebra will be invaluable for your applications of your knowledge of electronics to the digital world. These days the state of the art in electronics is digital signal processing. Since these logic circuits can become very large, involving dozens of gates, is it incumbent upon the technician or engineer to be able to design and construct the simplest circuit. Boolean algebra gives you this ability.

- A great advantage Boolean algebra gives the designer is the ability to design the logic in a formula, rather than use an experimental approach. The computer has greatly simplified this approach. Many CAD (computer aided design) programs exist today that are most helpful in simulating the logic functions for a particular application. Your study of this fascinating subject will prepare you well for your future in electronics.

## STUDENT QUIZ

Name ______________________________________________

1. The Boolean expression for the output of an AND gate is A + B.
   a. true
   b. false
2. An inverter gate has the output complemented from the input.
   a. true
   b. false
3. An OR gate has an output of 1 if any input is 1.
   a. true
   b. false
4. Boolean postulates and theorems seldom help to simplify a circuit.
   a. true
   b. false
5. The Boolean expression ABC is the output from an OR gate.
   a. true
   b. false
6. See figure 26-1 (a). Select the correct Boolean expression for the output of this circuit.
   a. $A + B + C$
   b. $(\overline{A + B})(\overline{B + C})$
   c. $\overline{A + B} + \overline{C}$
   d. $\overline{\overline{A}B}C$
7. See figure 26-1 (b). Select the correct Boolean expression for the output of this circuit.
   a. $A + B + C$
   b. $(\overline{A + B})(\overline{B + C})$
   c. $\overline{A + B} + \overline{C}$
   d. $\overline{\overline{A}B}C$
8. See figure 26-1 (c). Select the correct Boolean expression for the output of this circuit.
   a. $A + B + C$
   b. $(\overline{A + B})(\overline{B + C})$
   c. $\overline{A + B} + \overline{C}$
   d. $\overline{\overline{A}B}C$
9. See figure 26-1 (d). Select the correct Boolean expression for the output of this circuit.
   a. $A + B + C$
   b. $(\overline{A + B})(\overline{B + C})$
   c. $\overline{A + B} + \overline{C}$
   d. $\overline{\overline{A}B}C$

10. See Figure 26-1 (e). Select the correct Boolean expression for the output of this circuit.

a. A + B + C

b. $(\overline{\bar{A} + \bar{B}})(\overline{\bar{B} + \bar{C}})$

c. $\overline{\bar{A} + \bar{B}} + \bar{C}$

d. $\overline{\bar{\bar{A}}}BC$

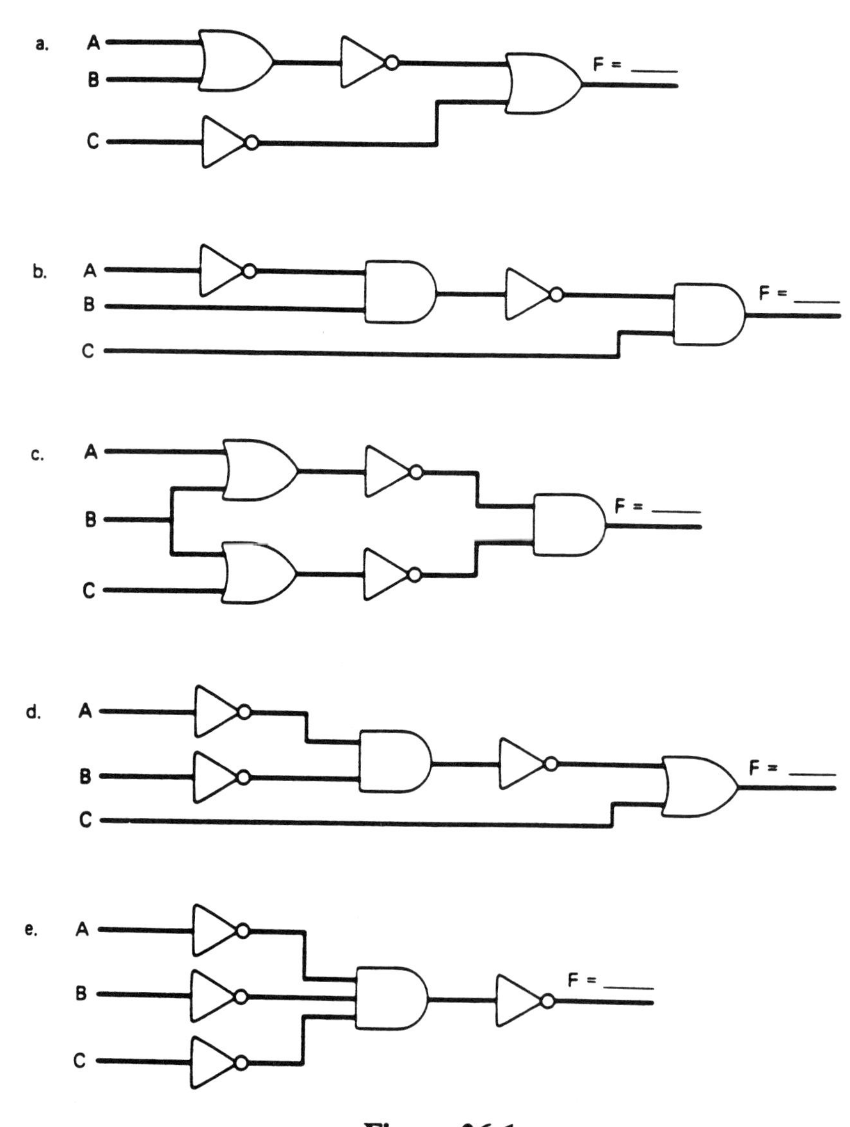


**Figure 26-1**

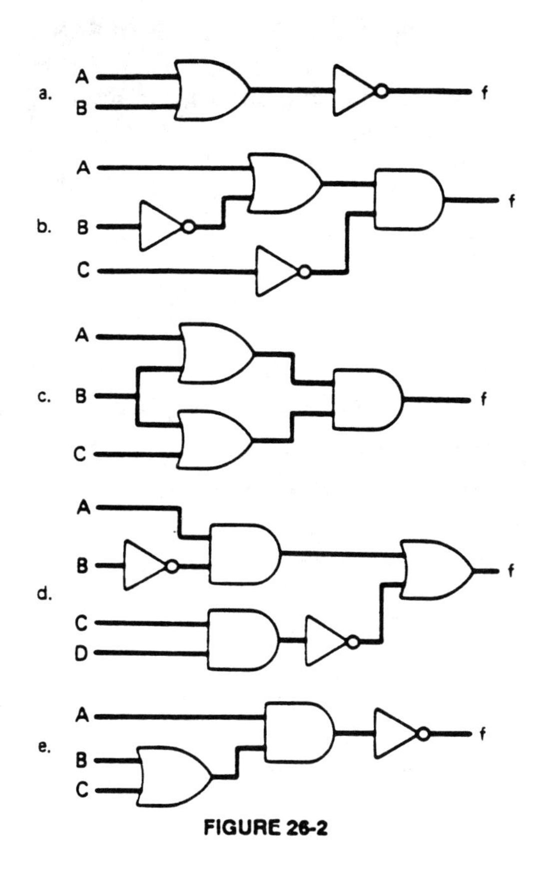

FIGURE 26-2

13. See figure 26-2. Given the following logic expression, select the logic circuit which will implement it correctly.

$$\overline{A(B + C)}$$

a. A
b. B
c. C
d. D
e. E

14. See figure 26-2. Given the following logic expression, select the logic circuit which will implement it correctly.

$$\overline{C}(A + \overline{B})$$

a. A
b. B
c. C
d. D
e. E

15. See figure 26-2. Given the following logic expression, select the logic circuit which will implement it correctly.

$$A\overline{B} + \overline{CD}$$

a. A
b. B
c. C
d. D
e. E

# KARNAUGH MAPS

CHAPTER 27

## TWO-VARIABLE EXPRESSION

- Designing complex digital circuits requires the process of simplification.
- One process of simplifying digital circuits is to use Boolean theorems and postulates.
- Another system of making digital circuits less complex is to use ***Karnaugh maps***.
- A Karnaugh map is a group of squares which represent the logic functions desired.
- These maps can have 2, 4, or more squares. The number of squares depends upon the quantity of variables. Two variables will require 4 squares, 3 variables will require 8, and so on.
- Take the data from a truth table and place it in the Karnaugh map squares.
- Allow only one change in variables as you go from square to square. The change must be from a complement to an uncomplement. As an example, start in the upper left square with $\overline{A}\,\overline{B}$. The next square to the right would be $A\overline{B}$
- Fill in the value for the bottom row.
- It is usual to only fill in the squares where the value is a 1. Look for adjacent pairs of squares where a variable does not change.
- OR the variables in the squares. The result is a Karnaugh map expression for the logic desired.
- If two horizontal or vertical squares have as an ORed expression $A\overline{B} + AB$, for example. Simplify it to A. Note: Only the common value A remains.

## THREE-VARIABLE EXPRESSION

- A truth table for a three-input circuit will have 8 possible combinations. The Karnaugh map will also have 8 squares.
- These squares may be in a 4 x 2 or 2 x 4 arrangement.
- A single square represents a three-variable term.
- Any two adjacent squares are combined to represent a two-variable term.
- A combination of four adjacent squares will represent one variable.
- Proceed to fill in the squares and OR the correct combinations together.
- Look for a wrap-around combination of squares at the ends of rows combining with the squares at the other end. These are valid combinations.

## FOUR-VARIABLE EXPRESSIONS

- The rule for a four-variable combination is similar in nature to those above.
  A four-variable term is represented by one square.
  A three-variable term is represented by 2 adjacent squares.
  A two-variable term is represented by 4 adjacent squares.
  A single term is represented by 8 adjacent squares.
- Use the same procedure to combine the terms. OR the combinations where the terms are the same.

# PRACTICAL COMMENTS

- Practice simplifying with Karnaugh maps. This system can be of importance to you, because it is a technique that works. You can actually visualize the Boolean expression. This will help you in your design problems. It is also important to become proficient in Boolean theorems which will, in combination with your Karnaugh map techniques, make you very adept at solving digital circuits.
- Karnaugh mapping techniques will be handy to use for designing such circuits as counters, multiplexers, flip-flop circuits, and many others.

**STUDENT QUIZ**

Name ______________________________________________

1. A Karnaugh map is useful for simplifying logic circuits.
   a. true
   b. false
2. A Boolean expression containing three terms will require a Karnaugh map with four squares.
   a. true
   b. false
3. Digital design dictates using the most complex circuits.
   a. true
   b. false
4. Karnaugh maps will have only one variable change from square to square, either vertically or horizontally.
   a. true
   b. false
5. Boolean simplification combined with Karnaugh mapping techniques are an effective design tool.
   a. true
   b. false
6. Use Karnaugh mapping to simplify the following expression:
   $\overline{A}\,\overline{B}C + \overline{A}\,\overline{B}\,\overline{C} + \overline{A}\,B\overline{C}$
   a. $\overline{A}$
   b. $\overline{A}B\overline{C} + C$
   c. $\overline{A}\,\overline{B} + \overline{A}\,\overline{C}$
   d. AB + BC
   e. AC
7. Use Karnaugh mapping to simplify the following expression:
   $A\overline{B}C + ABC$
   a. $\overline{A}$
   b. $\overline{A}B\overline{C} + C$
   c. $\overline{A}\,\overline{B} + \overline{A}\,\overline{C}$
   d. AB + BC
   e. AC
8. Use Karnaugh mapping to simplify the following expression:
   $\overline{A}\,\overline{B} + \overline{A}B$
   a. $\overline{A}$
   b. $\overline{A}B\overline{C} + C$
   c. $\overline{A}\,\overline{B} + \overline{A}\,\overline{C}$
   d. AB + BC
   e. AC
9. Use Karnaugh mapping to simplify the following expression:
   $\overline{A}BC + AB\overline{C} + ABC$
   a. $\overline{A}$
   b. $\overline{A}B\overline{C} + C$
   c. $\overline{A}\,\overline{B} + \overline{A}\,\overline{C}$
   d. AB + BC
   e. AC

10. Use Karnaugh mapping to simplify the following expression:

$\overline{A}\overline{B}C + \overline{A}B\overline{C} + ABC$ $\overline{A}\overline{B}C + \overline{A}B\overline{C} + ABC$

a. $\overline{A}$
b. $\overline{A}B\overline{C} + C$
c. $\overline{A}\,\overline{B} + \overline{A}\,\overline{C}$
d. $AB + BC$
e. $AC$

# INTRODUCTION TO STATISTICS

## )PULATION, RANGE, MEAN, MEDIAN, MODE

The entire set of data to be tested is a ***population***. We usually test a ***random sample*** of the ***population*** which means each item in the ***population*** has an equal chance of being selected.

A ***frequency distribution table*** tells how many items are in a study and how often each appears. These data are often shown using a ***histogram*** or ***bar graph***.

The ***range*** in a set of data is the difference between the extremes of data, between the largest (upper boundary number) and the smallest (lower boundary) numbers in the set.

The ***midpoint*** or ***median*** of a set of numbers is the middle value if there are an odd number of data items or is the average of the two middle values if there are an even number of data items. Half the data values are greater than the ***median*** and half are less.

The ***mean*** is the average or $\bar{x}$ of the set of data values. Add up all of the values and divide by the number of data items.

The ***mode*** is the data that occurs most frequently.

## ANDARD DEVIATION or SIGMA

The *standard deviation* (symbolized with lower case sigma, $\sigma$, or $\sigma_n$ on calculator ) is a measure of the distance any piece of data is from the ***mean***.

## NORMAL DISTRIBUTION CURVE

♦ In a normal distribution of values, the ***mean*** and the ***median*** and the ***mode*** occur at th same point.

♦ When data with normal distribution are graphed in a histogram and the high points form curve then:
  - the curve is bell shaped with only one peak, which is center
  - shapes are same on both sides of the peak, which means the data value are NO skewed (heavy on one side of mid-point)
  - ***mean*** and ***median*** and ***mode*** occur at the peak
  - 68% of values fall within ± one $\sigma$ of the ***mean***
  - 95% of values fall within ± two $\sigma$ of the ***mean***
  - 99.7% of values fall within ± three $\sigma$ of the ***mean***

♦ When analyzing data, the number of items sampled, ***n*** , can be very important. beautifully done analysis with a small number of data points might not be very meaningfu

♦ The size of the $\sigma$ is very important to any analysis. If you are manufacturing something, is important to keep the $\sigma$ within acceptable limits to be sure that the product mee manufacturing specifications.

## TATISTICAL ANALYSIS

Assume the following percentage scores were achieved in a class:

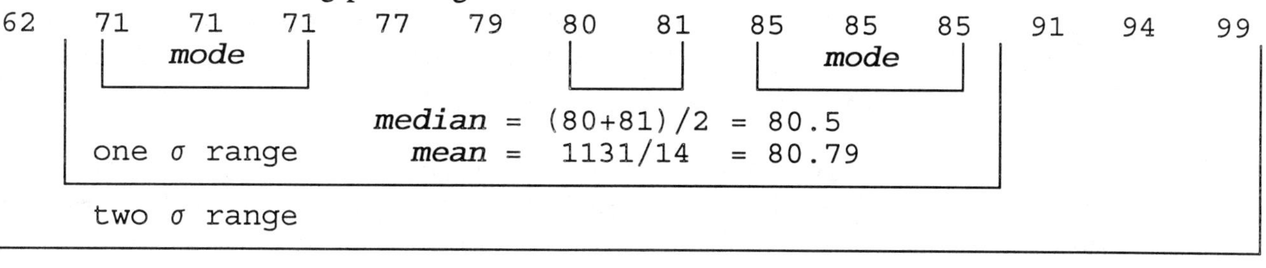


histogram gives:

| | | | | | | | | | |
|---|---|---|---|---|---|---|---|---|---|
| | X | | | | | X | | | |
| | X | | | | | X | | | |
| X | X | X | X | X | X | X | X | X | X |
| 62 | 71 | 77 | 79 | 80 | 81 | 85 | 91 | 94 | 99 |

better histogram
es:

| | | | | | | | | |
|---|---|---|---|---|---|---|---|---|
| | | X | | | X | | | |
| | | X | X | X | X | X | | |
| X | | X | X | X | X | X | X | |
| 60 | 65 | 70 | 75 | 80 | 85 | 90 | 95 | 100 |
| 1 | 0 | 3 | 2 | 2 | 3 | 2 | 1 | 0 |

intervals in sets of five: 60 = from 60 to 64.5

The extreme values are 99 and 62. The data range is 99 - 62 = 37.

The number of data items is 14, an even number.
The two middle numbers are 80 and 81. ***median*** = (80 + 81)/2 = 80.5

The most often given scores are 71 and 85, each of which is a ***mode***
This set is ***bimodal*** since there are two ***modes***

The ***mean*** or average value is the sum / number of data items or
(62 +71 +71 +71 +77 +79 +80 +81 +85 +85 +85 +91 +94 +99)/14
= 1131/14 = 80.79

The ***standard deviation*** or sigma or $\sigma$ or $\sigma_n$ is 9.72

The one $\sigma$ range is ( ***mean*** $\pm$ $\sigma$ ) and should include 68% of data.
(80.79 - 9.72 = 71.07 ) to (80.79 + 9.72 = 90.51) includes 10 out of the 14 or 71%, assuming 71 $\approx$ 71.07 is in the range.

The two $\sigma$ range is ( ***mean*** $\pm$ $2\sigma$ ) and should include 95% of data.
(71.07 - 9.72 = 61.35 ) to (90.51 + 9.72 = 100.23) includes all of the data.

# PRACTICAL COMMENTS

- A histogram or bar graph is a convenient way of showing how many items were in ea category. The histogram gives a visual idea of the relative sizes of the numbers withc having to read the individual numbers in a list or table. A line drawn across top of b gives an idea of how the different items are distributed.

- A histogram can be made using equal sized blocks (as on grid paper) or by using X (or a letter or symbol) with one letter for each block. Thc idea is to visually show the relati sizes of data items.

- If the number of data values is small, use a separate bar for each type of data. If the numb is larger, then group the data into similar sizes or shapes or numbers for each interval. sure to label the intervals you choose, because two people may choose different intervals a both can be correct.

- Before you start analyzing data, arrange them in order. Then determine the median a modes by just observing the arranged data.

- Statistical measurements can be calculated using formulas or almost automatically with ma algebraic calculators. You just have to change the calculator to statistical mode, enter data, then use the function keys to give you the mean and the sigma (standard deviatio This sigma is sometimes given on the calculator as $\sigma_n$ which is known as the populati sigma, which is separate from the $\sigma_{n-1}$ or Sx which is known as the sample sigma.

- Calculate your own grades in this class so far this semester.
  What grade do you think you have earned so far?

## STUDENT QUIZ

Name ________________________________________

1. The average of a set of data is called the:
   A. median B. mean C. mode D. sigma E. standard deviation

2. Which percentage of data items in a set should fall within one $\sigma$ range of the mean:
   A. 50% B. 60% C. 68% D. 70% E. 99%

3. In a normal distribution of values, the mean and the median and the mode occur at the same point. A. true B. false

Assume these percentage scores are given in a test:
58 63 63 68 72 72 76 78 84 84 84 92

4. The mean is:
   A. 9.93 B. 34 C. 58 D. 63 E. 68 F. 72 G. 74
   H. 74.5 I. 75.4 J. 76 K. 78 L. 84 M. 92 N. 9.93

5. The median is:
   A. 9.93 B. 34 C. 58 D. 63 E. 68 F. 72 G. 74
   H. 74.5 I. 75.4 J. 76 K. 78 L. 84 M. 92 N. 9.93

6. The mode is:
   A. 9.93 B. 34 C. 58 D. 63 E. 68 F. 72 G. 74
   H. 74.5 I. 75.4 J. 76 K. 78 L. 84 M. 92 N. 9.93

7. The standard deviation or sigma is:
   A. 9.93 B. 34 C. 58 D. 63 E. 68 F. 72 G. 74
   H. 74.5 I. 75.4 J. 76 K. 78 L. 84 M. 92 N. 9.93

8. The size of the range is:
   A. 9.93 B. 34 C. 58 D. 63 E. 68 F. 72 G. 74
   H. 74.5 I. 75.4 J. 76 K. 78 L. 84 M. 92 N. 9.93

9. The number of data points within one $\sigma$ is:
   A. 4 B. 5 C. 6 D. 7 E. 8 F. 9 G. 10

10. The number of data points within two $\sigma$ is :
A. 6 B. 7 C. 8 D. 9 E. 10 F. 11 G. 12

11. These data are distributed in a normal distribution.
A. true B. false

12. If the number of data items is odd, the median is the average of the two middle values, a
if the number is even then the median is the middle value itself.
A. true B. falsc

# ANSWERS TO STUDENT QUIZZES

## CHAPTER 1

| 1 | 2 | 3 | 4 | 5 | 6 | 7 | 8 | 9 | 10 | 11 | 12 | 13 | 14 | 15 |
|---|---|---|---|---|---|---|---|---|---|---|---|---|---|---|
| A | B | A | A | B | D | C | A | C | D | B | B | C | A | B |

| 16 | 17 | 18 | 19 | 20 |
|---|---|---|---|---|
| B | C | A | B | B |

## CHAPTER 2

| 1 | 2 | 3 | 4 | 5 | 6 | 7 | 8 | 9 | 10 | 11 | 12 | 13 | 14 | 15 |
|---|---|---|---|---|---|---|---|---|---|---|---|---|---|---|
| B | A | A | B | A | E | C | C | B | A | D | B | C | B | A |

## CHAPTER 3

| 1 | 2 | 3 | 4 | 5 | 6 | 7 | 8 | 9 | 10 | 11 | 12 | 13 | 14 | 15 |
|---|---|---|---|---|---|---|---|---|---|---|---|---|---|---|
| A | B | A | B | A | C | B | B | D | C | D | A | D | A | B |

## CHAPTER 4

| 1 | 2 | 3 | 4 | 5 | 6 | 7 | 8 | 9 | 10 | 11 | 12 | 13 | 14 | 15 |
|---|---|---|---|---|---|---|---|---|---|---|---|---|---|---|
| A | B | A | B | A | C | D | A | C | B | A | C | B | A | D |

## CHAPTER 5

| 1 | 2 | 3 | 4 | 5 | 6 | 7 | 8 | 9 | 10 | 11 | 12 | 13 | 14 | 15 |
|---|---|---|---|---|---|---|---|---|---|---|---|---|---|---|
| B | B | A | A | A | C | C | B | D | A | B | D | A | D | B |

## CHAPTER 6

| 1 | 2 | 3 | 4 | 5 | 6 | 7 | 8 | 9 | 10 | 11 | 12 | 13 | 14 | 15 |
|---|---|---|---|---|---|---|---|---|---|---|---|---|---|---|
| B | A | B | A | A | C | D | A | B | C | D | A | D | A | C |

## CHAPTER 7

| 1 | 2 | 3 | 4 | 5 | 6 | 7 | 8 | 9 | 10 | 11 | 12 | 13 | 14 | 15 |
|---|---|---|---|---|---|---|---|---|---|---|---|---|---|---|
| A | B | A | A | B | D | D | B | C | A | B | C | D | A | B |

## CHAPTER 8

| 1 | 2 | 3 | 4 | 5 | 6 | 7 | 8 | 9 | 10 | 11 | 12 | 13 | 14 | 15 |
|---|---|---|---|---|---|---|---|---|---|---|---|---|---|---|
| A | A | B | B | A | D | B | C | B | C | A | C | D | C | A |

## CHAPTER 9

| 1 | 2 | 3 | 4 | 5 | 6 | 7 | 8 | 9 | 10 | 11 | 12 | 13 | 14 | 15 |
|---|---|---|---|---|---|---|---|---|---|---|---|---|---|---|
| B | A | A | A | A | D | C | B | A | C | C | B | D | A | B |

## CHAPTER 10

| 1 | 2 | 3 | 4 | 5 | 6 | 7 | 8 | 9 | 10 |
|---|---|---|---|---|---|---|---|---|---|
| B | A | A | A | B | C | C | D | B | A |

## CHAPTER 11

| 1 | 2 | 3 | 4 | 5 | 6 | 7 | 8 | 9 | 10 | 11 | 12 | 13 | 14 | 15 |
|---|---|---|---|---|---|---|---|---|---|---|---|---|---|---|
| B | A | B | B | A | D | C | A | D | C | C | D | A | C | C |

## CHAPTER 12

| 1 | 2 | 3 | 4 | 5 | 6 | 7 | 8 | 9 | 10 | 11 | 12 | 13 | 14 | 15 |
|---|---|---|---|---|---|---|---|---|---|---|---|---|---|---|
| A | B | B | A | A | A | D | D | C | A | B | D | B | A | C |

## CHAPTER 13

| 1 | 2 | 3 | 4 | 5 | 6 | 7 | 8 | 9 | 10 |
|---|---|---|---|---|---|---|---|---|---|
| B | A | A | B | A | C | B | D | A | C |

## CHAPTER 14

| 1 | 2 | 3 | 4 | 5 | 6 | 7 | 8 | 9 | 10 | 11 | 12 | 13 | 14 | 15 |
|---|---|---|---|---|---|---|---|---|---|---|---|---|---|---|
| A | B | A | A | A | C | D | B | C | B | C | B | E | D | A |

## CHAPTER 15

| 1 | 2 | 3 | 4 | 5 | 6 | 7 | 8 | 9 | 10 |
|---|---|---|---|---|---|---|---|---|---|
| B | A | A | B | A | C | D | A | B | C |

## CHAPTER 16

| 1 | 2 | 3 | 4 | 5 | 6 | 7 | 8 | 9 | 10 |
|---|---|---|---|---|---|---|---|---|---|
| B | A | A | A | B | D | C | C | C | B |

## CHAPTER 17

| 1 | 2 | 3 | 4 | 5 | 6 | 7 | 8 | 9 | 10 | 11 | 12 | 13 | 14 | 15 |
|---|---|---|---|---|---|---|---|---|---|---|---|---|---|---|
| B | A | B | A | A | C | C | A | B | D | A | C | B | C | D |

## CHAPTER 18

| 1 | 2 | 3 | 4 | 5 | 6 | 7 | 8 | 9 | 10 |
|---|---|---|---|---|---|---|---|---|---|
| A | A | B | A | A | C | C | C | D | C |

## CHAPTER 19

| 1 | 2 | 3 | 4 | 5 | 6 | 7 | 8 | 9 | 10 | 11 | 12 | 13 | 14 | 15 |
|---|---|---|---|---|---|---|---|---|---|---|---|---|---|---|
| A | B | B | A | B | C | B | D | A | B | C | B | D | A | D |

## CHAPTER 20

| 1 | 2 | 3 | 4 | 5 | 6 | 7 | 8 | 9 | 10 | 11 | 12 | 13 | 14 | 15 |
|---|---|---|---|---|---|---|---|---|---|---|---|---|---|---|
| B | A | A | A | A | C | D | A | C | A | B | C | D | C | B |

## CHAPTER 21

| 1 | 2 | 3 | 4 | 5 | 6 | 7 | 8 | 9 | 10 |
|---|---|---|---|---|---|---|---|---|---|
| B | A | B | A | A | D | C | B | A | C |

## CHAPTER 22

| 1 | 2 | 3 | 4 | 5 | 6 | 7 | 8 | 9 | 10 | 11 | 12 | 13 | 14 | 15 |
|---|---|---|---|---|---|---|---|---|---|---|---|---|---|---|
| A | B | A | B | A | B | A | D | C | A | C | A | C | B | A |

## CHAPTER 23

| 1 | 2 | 3 | 4 | 5 | 6 | 7 | 8 | 9 | 10 |
|---|---|---|---|---|---|---|---|---|---|
| A | A | B | B | A | C | D | B | A | C |

## CHAPTER 24

| 1 | 2 | 3 | 4 | 5 | 6 | 7 | 8 | 9 | 10 |
|---|---|---|---|---|---|---|---|---|---|
| A | B | A | B | A | D | B | A | B | C |

## CHAPTER 25

| 1 | 2 | 3 | 4 | 5 | 6 | 7 | 8 | 9 | 10 |
|---|---|---|---|---|---|---|---|---|---|
| A | B | A | A | A | C | D | A | C | B |

## CHAPTER 26

| 1 | 2 | 3 | 4 | 5 | 6 | 7 | 8 | 9 | 10 | 11 | 12 | 13 | 14 | 15 |
|---|---|---|---|---|---|---|---|---|---|---|---|---|---|---|
| B | A | A | B | B | C | D | B | A | A | A | C | E | B | D |

## CHAPTER 27

| 1 | 2 | 3 | 4 | 5 | 6 | 7 | 8 | 9 | 10 |
|---|---|---|---|---|---|---|---|---|---|
| A | B | B | A | A | C | E | A | D | B |

## CHAPTER 28

| 1 | 2 | 3 | 4 | 5 | 6 | 7 | 8 | 9 | 10 | 11 | 12 |
|---|---|---|---|---|---|---|---|---|---|---|---|
| B | C | A | H | G | L | A | B | E | G | B | B |

# CHAPTER 1

***Problems 1-1***

1. In the number 8042: (a) 8 is the MSD; (b) 2 is the LSD; (c) 0 occupies the hundreds position; (d) 2 occupies the units position; (e) the 8 has a weight of 8000; (f) there are 4 tens.

2. In the number 67,403: (a) 6 is the MSD; (b) 3 is the LSD; (c) 7 occupies the thousands position; (d) 0 occupies the tens position; (e) the 7 has a weight of 7000; (f) there are 4 hundreds.

3. In the number 894,307; (a) 8 is the MSD; (b) 7 is the LSD; (c) 3 occupies the thousands position; (d) 9 occupies the tens position; (e) the 8 has a weight of 800,000 (f) there are 4 thousands.

***Problems 1-2***

1. (a) $\frac{3}{10} = 0.30$ (b) $\frac{16}{1000} = 0.016$ (c) $\frac{278}{100,000} = 0.00278$

   (d) $\frac{1763}{10,000} = 0.1763$ (e) $\frac{435}{1000} = 0.435$ (f) $\frac{2060}{10,000} = 0.206$

3. (a) $0.007 = \frac{7}{1000}$ (b) $0.0432 = \frac{432}{10.000}$ (c) $0.074= \frac{174}{1000}$

   (d) $0.000065 = \frac{65}{1,000,000}$ (e) $0.0016 = \frac{16}{10,000}$

   (f) $0.01234 = \frac{1234}{10,000}$

5. (a) 17 thousandths = $\frac{17}{1000} = 0.017$

   (b) 4 hundreths = $\frac{4}{100} = 0.04$

   (c) 460 ten-thousandths = $\frac{460}{10,000} = 0.0046$

   (d) 27 millionths = $\frac{27}{1,000,000} = 0.000027$

   (e) 1780 hundred-thousandths = $\frac{1780}{100,000} = 0.01780$

   (f) 65 thousandths = $\frac{65}{1000} = 0.0065$

7. In the number 0.001642:
   (a) the one appears in the thousandths place
   (b) the two appears in the millionths place
   (c) the six appears in the ten-thousandths place

9. In the number 0.508743:
   (a) the seven appears in the ten-thousandths place
   (b) the three appears in the millionths place
   (c) the zero appears in the hundredths place

11. (a) 0.006 is written six thousandths
    (b) 0.147 is written one hundred forty-seven thousandths
    (c) 0.00092 is written ninety-two hundred-thousandths
    (d) 0.000007 is written seven millionths
    (e) 0.0413 is written four hundred thirteen ten-thousandths
    (f) 0.0101 is written one hundred one ten-thousandths

***Problems 1-3***

1. (a) $7.14 = 7\frac{14}{100}$ (b) $50.02 = 50\frac{2}{100}$ (c) $710.143 = 710\frac{143}{1000}$

   (d) $9.099 = 9\frac{99}{1000}$ (e) $73.653 = 73\frac{653}{1000}$ (f) $207.7834 = 207\frac{7834}{10,000}$

   (g) $28.00736 = 28\frac{736}{100,000}$ (h) $8.0706 = 8\frac{706}{10,000}$

3. (a) $5\frac{68}{100} = 5.68$ (b) $25\frac{7}{1000} = 25.007$ (c) $7\frac{165}{10,000} = 7.0165$

   (d) $70\frac{4}{10} = 70.4$ (e) $473\frac{25}{1000} = 473.025$ (f) $80\frac{743}{100,000} = 80.00743$

   (g) $2475\frac{35}{1,000,000} = 2475.000035$ (h) $307\frac{8}{100,000} = 307.00008$

5. (a) Ninety-three and seven tenths = $93.7 = 93\frac{7}{10}$

   (b) Thirty and four hundredths = $30.04 = 30\frac{4}{100}$

   (c) Eleven and one ten-thousandths = $11.0001 = 11\frac{1}{10,000}$

   (d) Two hundred seventy-three and twenty-five hundred-thousandths = $273.00025 = 273\frac{25}{100,000}$

   (e) Seven hundred four and seven hundred four millionths = $704.000704 = 704\frac{704}{1,000,000}$

   (f) Three and thirty-three thousandths = $3.033 = 3\frac{33}{1000}$

***Problems 1-4***

| | *Original Number* | *Rounded to Nearest 10* | *Rounded to Nearest 100* | *Rounded to Nearest 1000* |
|---|---|---|---|---|
| 1. | 17 | 20 | | |
| 3. | 45 | 50 | | |
| 5. | 64 | 60 | | |
| 7. | 127 | 130 | | |
| 9. | 874 | 870 | | |
| 11. | 273 | 270 | 300 | |
| 13. | 356 | 360 | 400 | |
| 15. | 1377 | 1380 | 1400 | |
| 17. | 1407 | 1410 | 1400 | |
| 19. | 8706 | 8710 | 8700 | |
| 21. | 4817 | 4820 | 4800 | 5000 |
| 23. | 85468 | 85470 | 85500 | 85000 |
| 25. | 78673 | 78670 | 78700 | 79000 |
| 27. | 2784 | 2780 | 2800 | 3000 |
| 29. | 35486 | 35490 | 35500 | 36000 |
| 31. | 68448 | 68450 | 68400 | 68000 |
| 33. | 73654 | 73650 | 73700 | 74000 |

35. 465432 = 470,000 to the nearest ten-thousand

***Problems 1-5***

| | Original Number | Rounded to Nearest 100th | Rounded to Nearest 10th | Rounded to Nearest Unit | Rounded to Nearest 10 |
|---|---|---|---|---|---|
| 1. | 163.782 | 163.78 | 163.8 | 164 | 160 |
| 3. | 9.464 | 9.46 | 9.5 | 9 | 10 |
| 5. | 88.888 | 88.89 | 88.9 | 89 | 90 |
| 7. | 749.493 | 749.49 | 749.5 | 749 | 750 |
| 9. | 39.278 | 39.28 | 39.3 | 39 | 40 |
| 11. | 63.7478 | 63.75 | 63.7 | 64 | 60 |
| 13. | 478.6706 | 478.67 | 478.7 | 479 | 480 |
| 15. | 47.474 | 47.47 | 47.5 | 47 | 50 |
| 17. | 16.545 | 16.55 | 16.5 | 17 | 20 |
| 19. | 76.746 | 76.75 | 76.7 | 77 | 80 |

***Problems 1-6***

1. 12 - (+4) + (-3) = 12 - 4 - 3 = 5
3. 16 + 4 + (-6) = 16 + 4 - 6 = 14
5. -20 - (-4) - (-10) = -20 + 4 + 10 = -6
7. 30 + (-3) + 7 = 30 - 3 + 7 = 34
9. -10 - (+4) + (-14) = -10 - 4 - 14 = -28
11. 7 - (-7) - (+7) = 7 + 7 - 7 = 7
13. -5 - (-3) - (+12) = -5 + 3 -12 = -14
15. 14 + (-3) - (-7) = 14 - 3 + 7 = 18
17. 9 - (-4) - (-8) = 9 + 4 + 8 = 21
19. -20 - (+6) + (-7) = -20 - 6 - 7 = -33

***Problems 1-7***

1. 14 x (-3) = -42
3. -9 x (-6) = 54
5. -12 x 7 = -84
7. -45 x (-5) = 225
9. -310 x (-3) = 930
11. 28 ÷ (-4) = -7
13. -144 ÷ 6 = -24
15. -72 ÷ (-6) = 12
17. 512 ÷ (-16) = -32
19. -700 ÷ (-25) = 28

***Problems 1-8***

1. (-7) x 3 + 4 = -21 + 4 = -17
3. 6 x (-3) - 6 x (-3) = -18 - (-18) = 0
5. 6 + 3 x 5 - 4 = 6 + 15 - 4 = 17
7. 6 x 4 - 6 ÷ 2 = 24 - 3 = 21
9. (-7) - (-4) x 6 + 1 = -7 - (-24) + 1 = 18
11. -2 - (-2) x 4 - 5 x (-4) = -2 + 8 + 20 = 26
13. 16 ÷ (-4) - 7 x 3 + 4 = -4 - 21 + 4 = -21
15. -4 x (-5) - 3 + 15 ÷ (-3) = 20 - 3 - 5 = 12
17. (-35) ÷ 5 + 4 + 3 x 6 = -7 + 4 + 18 = 15
19. 7 x (-3) - 4 x (-5) + 6 x (-3) = -21 + 20 - 18 = -19

***Problems 1-9***

1. $-4 - (-2) + 6(6-3) = -4 + 2 + 6(3) = -4 + 2 + 18 = 16$

3. $7(-3) - [4(-7)] = -21 - (-28) = -21 + 28 = 7$

5. $(5+4)(-6+2) = (9)(-4) = -36$

7. $5 - 2 - 3[4(3+4)] - 2 = 3 - 3(28) - 2 = 3 - 84 - 2 = -83$

9. $4[24\div(-8+4)] + 7 \times 4 = 4(-6) + 28 = -24 + 28 = 4$

11. $15 - 4[(-3+6)(9-12)] = 15 - 4[(3)(-3)] = 15 - 4(-9) = 15 + 36 = 51$

13. $5[6-(-4)+(3-7)(2-6)] = 5[10+(-4)(-4)] = 5(10+16) = 5(26) = 130$

15. $4[-6-(-3)(4)] + 7 \times 3 = 4(-6+12) + 21 = 4(6) + 21 = 24 + 21 = 45$

17. $[9+(3\times-7)][18-(2\times3)] = (9-21)(18-6) = (-12)(12) = -144$

19. $14 - [(-3)\times4] + [3(4-5)(7-3)+6] = 14 - (-12) + [3(-1)(4)+6]$
$= 14 + 12 + [3(-4)+6] = 14 + 12 - 6 = 20$

# CHAPTER 2

***Problems 2-1***

1. $10^3 = 1000$  3. $10^0 = 1$  5. $10^6 = 1,000,000$

***Problems 2-2***

1. $10 = 10^1$  3. $100 = 10^3$  5. $100 = 10^2$

***Problems 2-3***

1. $10^{-2} = \frac{1}{100} = 0.01$  3. $10^{-3} = \frac{1}{1000} = 0.001$

5. $10^{-1} = \frac{1}{10} = 0.1$

***Problems 2-4***

1. $0.0001 = 10^{-4}$  3. $0.000001 = 10^{-6}$  5. $\frac{1}{10000} = 10^{-4}$  7. $\frac{1}{1000} = 10^{-3}$

***Problems 2-5***

1. $10^2 \times 10^6 = 10^8$  3. $10^2 \times 10^4 = 10^6$  5. $10^5 \times 10^{-1} = 10^4$

7. $10^3 \times 10^{-7} = 10^{-4}$  9. $10^{-1} \times 10^4 = 10^3$  11. $10^{-6} \times 10^{-4} = 10^{-10}$

13. $10^{-3} \times 10^{-9} = 10^{-12}$  15. $10^{-4} \times 10^{-5} = 10^{-9}$

***Problems 2-6***

1. $10^4$  3. $10^{-5}$  5. $10^8$  7. $10^{12}$

9. $10^7$  11. $10^{-3}$  13. $10^{-6}$  15. $10^{-7}$

17. $10^3$  19. $10^{-4}$

***Problems 2-7***

1. $10^{-7}$  3. $10^{-7}$  5. $10^{11}$  7. $10^{-10}$

9. $10^{-12}$  11. $10^8$  13. $10^{13}$  15. $10^{-4}$

17. $10^{-36}$

***Problems 2-8***

| | | (a) | (b) | (c) |
|---|---|---|---|---|
| 1. | 475 = | $4.75 \times 10^2$ = | $0.0475 \times 10^4$ = | $0.000475 \times 10^6$ |
| 3. | 65.9 = | $0.0659 \times 10^2$ = | $0.00659 \times 10^4$ = | $0.0000659 \times 10^6$ |
| 5. | 93400 = | $924 \times 10^2$ = | $9.34 \times 10^4$ = | $0.0934 \times 10^6$ |
| 7. | 4780 = | $47.8 \times 10^2$ = | $0.478 \times 10^4$ = | $0.00478 \times 10^6$ |
| 9. | 0.00465 = | $0.465 \times 10^{-2}$ = | $46.5 \times 10^{-4}$ = | $4650 \times 10^{-6}$ |
| 11. | 0.00000555 = | $0.000555 \times 10^{-2}$ = | $0.0555 \times 10^{-4}$ = | $5.55 \times 10^{-6}$ |
| 13. | 0.000673 = | $0.0673 \times 10^{-2}$ = | $6.73 \times 10^{-4}$ = | $673 \times 10^{-6}$ |
| 15. | 0.0000000108 = | $0.00000108 \times 10^{-2}$ = | $0.000108 \times 10^{-4}$ = | $0.0108 \times 10^{-6}$ |
| 17. | 325 = | $0.325 \times 10^3$ = | $325000 \times 10^{-3}$ | |

19. 4.75 = $0.00475 \times 10^{3}$ = $4750 \times 10^{-3}$

***Problems 2-9***

1. $2.76 \times 10^{4}$ 3. $4.78 \times 10^{1}$ 5. $1.77 \times 10^{6}$
7. $2.73 \times 10^{2}$ 9. $1.73 \times 10^{5}$ 11. $7.89 \times 10^{1}$
13. $1.67 \times 10^{6}$ 15. $5.79 \times 10^{4}$ 17. $8.10 \times 10^{4}$
19. $4.71 \times 10^{5}$

***Problems 2-10***

1. $4.78 \times 10^{-3}$ 3. $7.47 \times 10^{1}$ 5. $1.64 \times 10^{-2}$
7. $2.78 \times 10^{-6}$ 9. $1.71 \times 10^{-2}$ 11. $8.96 \times 10^{-4}$
13. $4.54 \times 10^{-3}$ 15. $5.01 \times 10^{-5}$ 17. $2.55 \times 10^{-1}$
19. $4.00 \times 10^{-2}$

***Problems 2-11***

1. $2.13 \times 10^{4}$ 3. $1.28 \times 10^{-1}$ 5. $1.05 \times 10^{4}$
7. $6.13 \times 10^{0}$ 9. $2.99 \times 10^{0}$ 11. $1.31 \times 10^{-3}$
13. $1.23 \times 10^{10}$ 15. $2.18 \times 10^{-6}$ 17. $6.70 \times 10^{0}$
19. $1.82 \times 10^{-2}$ 21. $9.41 \times 10^{-5}$ 23. $2.59 \times 10^{6}$

***Problems 2-12***

1. $1.10 \times 10^{4}$ 3. $2.00 \times 10^{-1}$ 5. $3.86 \times 10^{3}$
7. $4.82 \times 10^{5}$ 9. $2.88 \times 10^{-1}$ 11. $3.87 \times 10^{-2}$
13. $9.93 \times 10^{4}$ 15. $4.26 \times 10^{5}$

***Problems 2-13***

1. $1.06 \times 10^{2}$ 3. $1.27 \times 10^{-4}$ 5. $3.26 \times 10^{-3}$ 7. $1.67 \times 10^{-3}$
9. $3.91 \times 10^{-5}$ 11. $5.16 \times 10^{-6}$ 13. $1.10 \times 10^{-4}$ 15. $2.56 \times 10^{3}$
17. $8.46 \times 10^{1}$ 19. $3.07 \times 10^{4}$ 21. $2.44 \times 10^{3}$ 23. $1.39 \times 10^{0}$
25. $1.74 \times 10^{0}$ 27. $1.09 \times 10^{0}$ 29. $2.53 \times 10^{0}$ 31. $1.03 \times 10^{-2}$
33. $6.75 \times 10^{-3}$

***Problems 2-14***

1. $10^{6}$ 3. $10^{10}$ 5. $10^{-12}$ 7. $10^{-8}$ 9. $10^{-8}$

***Problems 2-15***

1. $10^3$ 3. $10^{-2}$ 5. $10^{-2}$ 7. $10^5$
9. $10^3$

***Problems 2-16***

1. $6.29 \times 10^3$ 3. $3.17 \times 10^6$ 5. $2.26 \times 10^{-3}$ 7. $7.62 \times 10^{-5}$
9. $2.03 \times 10^9$ 11. $9.00 \times 10^{10}$ 13. $8.76 \times 10^{-8}$ 15. $8.15 \times 10^{-9}$
17. $6.76 \times 10^0$ 19. $2.17 \times 10^0$

***Problems 2-17***

1. $9.64 \times 10^0$ 3. $1.54 \times 10^1$ 5. $8.40 \times 10^{-2}$ 7. $1.02 \times 10^{-2}$
9. $5.20 \times 10^3$ 11. $6.71 \times 10^3$ 13. $3.04 \times 10^{-3}$ 15. $5.88 \times 10^0$
17. $1..07 \times 10^{-2}$ 19. $7.96 \times 10^3$ 21. $1.58 \times 10^4$

# CHAPTER 3

***Problems 3-1***

1. $56 \times 10^3 = 0.056 \times 10^6$
3. $220 \times 10^3 = 0.22 \times 10^6$
5. $390 \times 10^3 = 0.39 \times 10^6$
7. $180 \times 10^3 = 0.18 \times 10^6$
9. $43 \times 10^3 = 0.043 \times 10^6$
11. $0.22 \times 10^{-3} = 220 \times 10^{-6}$
13. $2.13 \times 10^{-3} = 2130 \times 10^{-6}$
15. $0.0556 \times 10^{-3} = 55.6 \times 10^{-6}$
17. $0.122 \times 10^{-3} = 122 \times 10^{-6}$
19. $0.256 \times 10^{-3} = 256 \times 10^{-6}$
21. $667 \times 10^{-9} = 667{,}000 \times 10^{-12}$
23. $17.9 \times 10^{-9} = 17{,}900 \times 10^{-12}$
25. $6.74 \times 10^{-9} = 6740 \times 10^{-12}$
27. $0.0177 \times 10^{-9} = 17.7 \times 10^{-12}$
29. $70 \times 10^{-9} = 70{,}000 \times 10^{-12}$
31. $7.3 = 7300 \times 10^{-3} = 0.0073 \times 10^3$
33. $5.67 = 5670 \times 10^{-3} = 0.00567 \times 10^3$
35. $0.0178 = 17.8 \times 10^{-3} = 0.0000178 \times 10^3$
37. $78.3 = 78{,}300 \times 10^{-3} = 0.0783 \times 10^3$
39. $845 = 845{,}000 \times 10^{-3} = 0.845 \times 10^3$

***Problems 3-2***

1. 0.26 mA = 260 μA
3. 0.632 mS = 632 μS
5. 7.63 kΩ = 0.00763 MΩ
7. 17.3 mA = 17,300 μA
9. 713 kΩ = 0.713 MΩ
11. 5630 kHz = 5.63 MHz
13. 2000 kΩ = 2 MΩ
15. 0.237 mS = 237 μS
17. 0.3 μF = 300 nF
19. 0.062 mA = 62 μA

***Problems 3-3***

1. 0.8 A
3. 0.0025 S
5. 33,000 Ω
7. 470 Ω
9. 12,500 Hz
11. 0.0001 F
13. 0.00025 A
15. 0.0009 S
17. 750,000 Ω
19. 0.03 A

***Problems 3-4***

1. 0.26 mA = 260 μA
3. 0.00632 mS = 6.32 μS
5. 56.2 mS = 56,200 μS
7. 0.613 mS = 613 μS
9. 7.63 kΩ = 0.00763 MΩ
11. 470 kΩ = 0.47 MΩ
13. 127 kHz = 0.127 MHz
15. 713 kΩ = 0.713 MΩ
17. 46,300 μA = 0.0463 μA
19. 0.00005 μF = 50 pF
21. 3200 Ω = 0.0032 MΩ
23. 0.27 MΩ = 270,000 Ω
25. 0.000403 S = 0.403 mS
27. 1,030,000 Hz = 1030 kHz
29. 1430 pF = 0.00143 μF
31. 5.5 mS = 5500 μS
33. 0.00106 μF = 1060 pF
35. 463,000 Ω = 0.463 MΩ
37. 96.3 kHz = 0.0963 MHz
39. 7.8 mA = 7800 μA
41. 0.176 V = 176,000 μV
43. 1730 mW = 0.00173 kW
45. 25 mH = 25,000 μH
47. 0.173 S = 173,000 μS
49. 2.5 μF = 2500 nF

***Problems 3-5***

1. 2.04 mA = 2040 μA
3. 0.0957 mA = 95.7 μA
5. 0.167 mS = 167 μS
7. 0.0582 mS = 58.2 μS
9. 1.42 mS = 1420 μS
11. 964 Ω = 0.964 kΩ

13. 68700 Ω = 68.7 kΩ = 0.0687 MΩ

15. 6850 Ω = 6.85 kΩ = 0.00685 MΩ

17. 115 Ω = 0.115 kΩ

19. 4690 Ω = 4.69 kΩ

21. 2.30 mS = 2300 μS

23. 0.273 mA = 273 μA

25. 70,000 Ω = 70 kΩ

27. 0.042 mA = 41.7 μA

29. 9.52 V = 9520 mV

31. 4.48 V = 4480 mV

33. 84900 Ω = 84.9 kΩ

35. 318 Ω = 0.318 kΩ

37. 0.0049 μF = 4.91 nF

39. 919 Hz = 0.919 kHz

***Problems 3-6***

1. 1 cm = 0.3937 in: 100 cm x 0.3937 = 39.37 in

3. 40 cm = 40 x 0.3937 = 15.75 in

5. 1 m = 39.37 in: 2.3 m = 2.3 x 39.37 = 90.55 in
   1 m = 1.094 yd: 2.3 m = 2.3 x 1.094 = 2.516 yd

7. 10 m = 10.94 yd

9. 10,000 m = 10,940 yd
   1 km = 0.6214 mi: 10 km = 10 x 0.6214 = 6.214 mi

11. 1 in = 2.54 cm: 10 in = 10 x 2.54 = 25.4 cm

13. 35.6 in = 90.42 cm

15. 1 yd = 0.9141 m or 0.0009141 km
    880 yds = 880 x 0.9141 = 804.4 m or 0.8044 km

17. 1 mi/hr = 1.609 km/h: 55 mi/h x 1.609 = 88.50 km/h

19. 100 mi/h = 160.9 km/h

21. 500 m/s x 3.281 ft/s = 1640 ft/s

23. 100 m/s = 328.1 ft/s

25. 1 g = 0.03527 oz: 60 g x 0.03527 = 2.116 oz

27. 150 g = 5.291 oz

29. 800 g = 28.22 oz
    1 kg = 2.205 lb: 800 g = 0.8 kg: 0.8 kg x 2.205 = 1.764 lb

31. 5 kg = 176.4 oz or 11.03 lb

33. 1 lb. = 0.4536 kg: 1.35 lb x 0.4536 = 0.6124 kg

35. 1 oz = 28.35 g: 4 oz x 28.35 = 113.4 g

37. 0.275 oz = 7.796 g

# CHAPTER 4

***Problems 4-1***

1. 7   3. 10   5. 13   7. 23   9. 38

11. 56   13. 101   15. 119   17. 240   19. 195

21. $101_2$   23. $1100_2$   25. $10101_2$   27. $101101_2$   29. $111011_2$

31. $1000100_2$   33. $1100000_2$   35. $10000111_2$   37. $11010010_2$

***Problems 4-2***

1. $12_{10}$   3. $63_{10}$   5. $190_{10}$   7. $2174_{10}$   9. $5068_{10}$

11. $24_8$   13. $120_8$   15. $550_8$   17. $2611_8$   19. $7315_8$

***Problems 4-3***

1. $26_{10}$   3. $76_{10}$   5. $512_{10}$   7. $4522_{10}$   9. $41{,}003_{10}$

11. $16_{16}$   13. $61_{16}$   15. $200_{16}$   17. $A8C_{16}$   19. $17BB_{16}$

***Problems 4-4***

1. $312_8 = CA_{16}$   3. $233_8 = 9B_{16}$   5. $201_8 = 81_{16}$   7. $6311_8 = CC9_{16}$

9. $5303_8 = AC3_{16}$   11. $1111_2$   13. $100100_2$   15. $10000000_2$

17. $101110000_2$   19. $1000000000_2$   21. $1100_2$   23. $100100_2$

25. $1001111_2$   27. $11100011_2$   29. $110011010_2$

***Problems 4-5***

1. $111101_2 = 61_{10} = 3D_{16}$   3. $1100000_2 = 96_{10} = 60_{16}$

5. $101010_2 = 52_8 = 42_{10}$   7. $110100111_2 = 647_8 = 423_{10}$

9. $46_8 = 38_{10} = 26_{16}$   11. $705_8 = 453_{10} = 1C5_{16}$

13. $1010_2 = 12_8 = A_{16}$   15. $100100010_2 = 442_8 = 122_{16}$

***Problems 4-6***

1. $357_8$   3. $775_8$   5. $776_8$   7. $1423_8$   9. $1107_8$

***Problems 4-7***

1. $ED_{16}$   3. $CA5_{16}$   5. $D56_{16}$   7. $179D_{16}$   9. $1DA6C_{16}$

***Problems 4-8***

1. $10101_2$   3. $110011_2$   5. $101001_2$   7. $1100010_2$   9. $1101001_2$

***Problems 4-9***

1. $45_8$   3. $42_8$   5. $245_8$   7. $337_8$   9. $1515_8$

11. $1A_{16}$   13. $3E1_{16}$   15. $5F1_{16}$   17. $AFD_{16}$   19. $FDC_{16}$

21. $101_2$   23. $10_2$   25. $10_2$   27. $11_2$   29. $10110_2$

***Problems 4-10***

1. 1010
$$\begin{array}{rl} 0101 & \text{1's comp} \\ +\underline{\quad 1} & \\ 0110 & \text{2's comp} \end{array}$$

3. 101101
$$\begin{array}{rl} 010010 & \text{1's comp} \\ +\underline{\quad\quad 1} & \\ 010011 & \text{2's comp} \end{array}$$

5. 64
$$\begin{array}{rl} 77 & \\ -\underline{64} & \\ 13 & \text{7's comp} \\ +\underline{\ 1} & \\ 14 & \text{8's comp} \end{array}$$

7. 635
$$\begin{array}{rl} 777 & \\ -\underline{635} & \\ 142 & \text{7's comp} \\ +\underline{\quad 1} & \\ 143 & \text{8's comp} \end{array}$$

```
9.  A3     FF
          -A3
           5C    15's comp
          + 1
           5D    16's comp
```

```
11. C1E    FFF
          -C1E
           3E1   15's comp
          +  1
           3E2   16's comp
```

```
13.  11001    11001
    -01100   +10011    1's comp
             101100
              └--->1   eac
              01101    difference
```

```
15.  101101    101101
    -001010   +110101    1's comp
              1100010
               └--->1    eac
               100011    difference
```

```
17.  10110    10110
    -01011   +10101    2's comp
             1̸01011    difference
```

```
19.  1011001     1011001
    -0010110    +1101010    2's comp
                1̸1000011    difference
```

```
21.  63    63
    -25   +53    8's comp
          1̸36    difference
```

```
23.   626    626
     -377   +401    8's comp
            1̸227    difference
```

```
25.  A2    A2
    -8C   +74    16's comp
          1̸16    difference
```

```
27.  6A2    6A2
    -4FF   +B01    16's comp
           1̸1A3    difference
```

# CHAPTER 5

***Problems 5-1***

1. Two terms, binomial expression
3. Three terms, trinomial expression
5. Two terms, binomial expression
7. Two terms, binomial expression
9. Three terms, trinomial expression
11. $3a - 4b + 2c$: The coefficients are 3, 4, and 2. The literal numbers are a, b, and c.
13. $\frac{2x}{3} + 7y$: The coefficients are $\frac{2}{3}$ and 7. The literal numbers are x and y.
15. $\frac{5}{2a} - \frac{4}{b}$: The coefficients are $\frac{5}{2}$ and -4. The literal numbers are a and b.

***Problems 5-2***

1. 5.29 3. 54.8 5. 256 7. 512 9. 46.7
11. 166 13. 16.0 15. 64.0 17. 256

***Problems 5-3***

1. $ac = 5\cdot6 = 30.0$
3. $a^2c = 5^2\cdot6 = 25\cdot6 = 150$
5. $(ac)^2 = (5\cdot6)^2 = 30^2 = 900$
7. $2(ab)^2 = 2(5\cdot3)^2 = 2(15)^2 = 2\cdot225 = 450$
9. $a^2b^2 - 3c^2 = 5^2\cdot3^2 - 3\cdot6^2 = 25\cdot9 - 3\cdot36 = 225 - 108 = 117$
11. $ac^2 - 2abc = 5\cdot6^2 - 2\cdot5\cdot3\cdot6 = 180 - 180 = 0$
13. $(a^2b - 2c^2)^2 = (5^2\cdot3 - 2\cdot6^2)^2 = (75 - 72)^2 = 3^2 = 9.00$
15. $3a^2b^3c - 2a^3b^2c^2 = 3\cdot5^2\cdot3^3\cdot6 - 2\cdot5^3\cdot3^2\cdot6^2 = 3\cdot25\cdot27\cdot6 - 2\cdot125\cdot9\cdot36$ $= 12{,}150 - 81{,}000 = -68{,}850 = -6.89 \times 10^4$
17. $3ab^3c + 4a^2bc^3 - 2a^4bc^2 = 3\cdot5\cdot3^3\cdot6 + 4\cdot5^2\cdot3\cdot6^3 - 2\cdot5^4\cdot3\cdot6^2$ $= 2.43\times10^3 + 6.48\times10^4 - 1.35\times10^5 = -6.78 \times 10^4$
19. $(3a^2c^4 - 2a^3b^2c)^3 = (3\cdot5^2\cdot6^4 - 2\cdot5^3\cdot3^2\cdot6)^3 = (9.72\times10^4 - 1.35\times10^4)^3$ $= (8.37\times10^4)^3 = 5.86 \times 10^{14}$

***Problems 5-4***

1. 5.48 3. 6.32 5. 14.1 7. 32.2 9. 44.7

***Problems 5-5***

1. $\sqrt{xy} = \sqrt{4\cdot5} = \sqrt{20} = 4.47$
3. $\sqrt{y^2} = 4.00$
5. $\sqrt{6xz^2} = \sqrt{450} = 2.12 \times 10^1 = 21.2$
7. $\sqrt{3x^2} = \sqrt{57} = 7.55$
9. $\sqrt{6.4y^3 - z^2} = \sqrt{384.6} = 1.96 \times 10^1 = 19.6$
11. $\sqrt{x^2y^2} + x = \sqrt{400} + 3 = 23.0$

13. $\sqrt{x^3} + 2z = \sqrt{5.20} + 10 = 1.52 \times 10^1 = 15.2$

15. $\sqrt{x^3z^2 + 3x^4y^2} = \sqrt{675 + 3888} = 6.75 \times 10^1 = 67.5$

17. $\sqrt{3y^3z^3 - 5x^3z^3} = \sqrt{24000 - 16875} = 8.44 \times 10^1 = 84.4$

19.

$\sqrt{x^3y^4z^2 - 2x^2y^3z} - z = \sqrt{4 \cdot 27 \cdot 256 \cdot 5 - 2 \cdot 9 \cdot 64 \cdot 5} - 5 = \sqrt{691200 - 5760} = 823$

21. $\sqrt{5.6x^4y^3z^4 - 3x^2y^4z^3} - 4z^2 = \sqrt{18144000 - 864000} - 500 = 3.66 \times 10^3 = 3660$

***Problems 5-6***

1. $E = 660\ \mu A \cdot 6.8\ k\Omega = 4.49\ V$

3. $E = 3.7\ mA \cdot 18\ k\Omega = 66.6\ V$

5. $E = 50\ mA \cdot 2.7\ k\Omega = 135\ V$

7. $I = 430\ mA \cdot 560\ \Omega = 241\ V$

9. $P = (7.5\ mA)^2 \cdot 680\ \Omega = 38.3\ mW$

11. $P = (36.5\ mA)^2 \cdot 3.3\ k\Omega = 4.4\ W$

13. $P = (1.2\ A)^2 \cdot 5\ \Omega = 7.20\ W$

15. $P = (23.5\ mA)^2 \cdot 810\ \Omega = 0.447\ mW$

17. $I = \sqrt{\dfrac{65.4\ mW}{5.6\ k\Omega}} = 3.42\ mA$

19. $I = \sqrt{\dfrac{50\ mW}{18\ k\Omega}} = 1.67\ mA$

21. $I = \sqrt{\dfrac{4.6\ W}{10\ \Omega}} = 678\ mA$

23. $\sqrt{\dfrac{650\ \mu W}{56\ \Omega}} = 3.41\ mA$

25. $P = \sqrt{8.3\ mW \cdot 27\ k\Omega} = 15.0\ V$

27. $P = \sqrt{130\ mW \cdot 750\ \Omega} = 9.87\ V$

29. $P = \sqrt{18\ mW \cdot 6.8\ k\Omega} = 11.1\ V$

31. $P = \sqrt{100\ mW \cdot 100\ \Omega} = 3.16\ V$

33. $P = \dfrac{(16\ V)^2}{6.8\ k\Omega} = 37.6\ mW$

35. $P = \dfrac{(13.3\ V)^2}{33\ k\Omega} = 5.36\ mW$

37. $P = \dfrac{(18.2\ V)^2}{10\ k\Omega} = 33.1\ mW$

39. $P = \dfrac{(150\ V)^2}{4.7\ k\Omega} = 4.79\ W$

***Problems 5-7***

1. $X_C = \dfrac{1}{2\pi \cdot 5\ kHz \cdot 100\ nF} = 318\ \Omega$

3. $X_C = \dfrac{1}{2\pi \cdot 200\ Hz \cdot 200\ nF} = 3.98\ k\Omega$

5. $X_C = \dfrac{1}{2\pi \bullet 6.3\ kHz \bullet 50\ nF} = 12.6\ \Omega$

7. $X_C = \dfrac{1}{2\pi \cdot 65.4 kHz \cdot 250 pF} = 9.73\ k\Omega$

9. $X_C = \dfrac{1}{2\pi \cdot 130\ kHz \cdot 750\ pF} = 1.63\ k\Omega$

11. $f_r = \dfrac{1}{6.28\sqrt{80\ mH \cdot 150\ nF}} = 1.45\ kHz$

13. $f_r = \dfrac{1}{6.28\sqrt{150\ \mu F \cdot 50\ nF}} = 58.1\ kHz$

15. $f_r = \dfrac{1}{6.28\sqrt{50\ mH \cdot 250\ nF}} = 1.42 kHz$

17. $f_r = \dfrac{1}{6.28\sqrt{400\ mH \cdot 5\ nF}} = 3.56\ kHz$

19. $f_r = \dfrac{1}{6.28\sqrt{0.25\ mH \cdot 25\ \mu F}} = 63.7\ Hz$

21. $Z = \sqrt{(120\ k\Omega)^2 + (80\ k\Omega)^2} = 144\ k\Omega$

23. $Z = \sqrt{(200\ \Omega)^2 + (100\ \Omega)^2} = 224\ \Omega$

25. $Z = \sqrt{(33\ k\Omega)^2 + (20\ k\Omega)^2} = 38.6\ k\Omega$

27. $Z = \sqrt{(4.7\ k\Omega)^2 + (7.35\ k\Omega)^2} = 8.72\ k\Omega$

29. $Z = \sqrt{(2.2\ M\Omega)^2 + (6.75\ M\Omega)^2} = 7.10\ M\Omega$

# CHAPTER 6

***Problems 6-1***

1. $18 = 2 \cdot 3 \cdot 3$
3. $44 = 2 \cdot 2 \cdot 11$
5. $63 = 3 \cdot 3 \cdot 7$
7. $92 = 2 \cdot 2 \cdot 23$
9. $231 = 3 \cdot 7 \cdot 11$
11. $56 = 2 \cdot 2 \cdot 2 \cdot 7$
13. $147 = 3 \cdot 7 \cdot 7$
15. $84 = 2 \cdot 2 \cdot 3 \cdot 7$
17. $210 = 2 \cdot 3 \cdot 5 \cdot 7$
19. $455 = 5 \cdot 7 \cdot 13$

***Problems 6-2***

1. $\frac{\cancel{3}}{\cancel{3} \cdot 7} = \frac{1}{7}$

3. $\frac{5}{2 \cdot 2 \cdot 2} = \frac{5}{8}$

5. $\frac{11 \cdot \cancel{2}}{4 \cdot 4 \cdot \cancel{2}} = \frac{11}{16}$

7. $\frac{\cancel{2} \cdot 2 \cdot \cancel{3} \cdot \cancel{3}}{\cancel{3} \cdot \cancel{3} \cdot 5} = \frac{2}{5}$

9. $\frac{3 \cdot 7}{4 \cdot 2 \cdot 5} = \frac{21}{40}$

11. $\frac{\cancel{2} \cdot \cancel{11} \cdot 3}{\cancel{2} \cdot \cancel{11} \cdot 5} = \frac{3}{5}$

13. $\frac{2 \cdot 7 \cdot \cancel{13}}{3 \cdot 5 \cdot \cancel{13}} = \frac{14}{15}$

15. $\frac{\cancel{3} \cdot \cancel{7} \cdot 13}{\cancel{3} \cdot 3 \cdot \cancel{7} \cdot 7} = \frac{13}{21}$

17. $\frac{7 \cdot 7 \cdot \cancel{2} \cdot \cancel{3}}{5 \cdot 11 \cdot \cancel{2} \cdot \cancel{3}} = \frac{49}{55}$

19. $\frac{\cancel{5} \cdot \cancel{2} \cdot \cancel{19}}{\cancel{5} \cdot \cancel{2} \cdot 3 \cdot \cancel{19}} = \frac{1}{3}$

***Problems 6-3***

1. $\frac{1}{2} \times \frac{1}{3} = \frac{1}{6}$

3. $\frac{3}{8} \times \frac{2}{9} = \frac{\cancel{2} \cdot \cancel{3}}{\cancel{2} \cdot 2 \cdot 2 \cdot \cancel{3} \cdot 3} = \frac{1}{12}$

5. $\frac{9}{16} \times \frac{8}{15} = \frac{\cancel{2} \cdot \cancel{2} \cdot \cancel{2} \cdot \cancel{3} \cdot 3}{\cancel{2} \cdot \cancel{2} \cdot \cancel{2} \cdot 2 \cdot \cancel{3} \cdot 5} = \frac{3}{10}$

7. $\frac{22}{33} \times \frac{15}{28} = \frac{\cancel{2} \cdot \cancel{3} \cdot 5 \cdot \cancel{11}}{\cancel{2} \cdot 2 \cdot \cancel{3} \cdot 7 \cdot \cancel{11}} = \frac{5}{14}$

9. $\frac{3}{4} \times \frac{7}{8} = \frac{21}{32}$

11. $\frac{3}{8} \times \frac{4}{9} \times \frac{2}{3} = \frac{2 \cdot 2 \cdot 2 \cdot 3}{2 \cdot 2 \cdot 2 \cdot 3 \cdot 3 \cdot 3} = \frac{1}{9}$

13. $\frac{1}{3} \times \frac{1}{4} \times \frac{2}{5} = \frac{2}{\cancel{2} \cdot 2 \cdot 3 \cdot 5} = \frac{1}{30}$

15. $\frac{1}{5} \times \frac{4}{7} \times \frac{1}{2} = \frac{\cancel{2} \cdot 2}{\cancel{2} \cdot 5 \cdot 7} = \frac{2}{35}$

17. $\frac{6}{35} \times \frac{14}{15} \times \frac{5}{12} = \frac{\cancel{2} \cdot \cancel{2} \cdot \cancel{3} \cdot \cancel{5} \cdot \cancel{7}}{\cancel{2} \cdot \cancel{2} \cdot \cancel{3} \cdot 3 \cdot \cancel{5} \cdot 5 \cdot \cancel{7}} = \frac{1}{15}$

19. $\frac{8}{15} \times \frac{5}{16} \times \frac{22}{25} = \frac{\cancel{2} \cdot \cancel{2} \cdot \cancel{2} \cdot \cancel{2} \cdot \cancel{5} \cdot 11}{\cancel{2} \cdot \cancel{2} \cdot \cancel{2} \cdot \cancel{2} \cdot 3 \cdot \cancel{5} \cdot 5 \cdot 5} = \frac{11}{75}$

***Problems 6-4***

1. $\frac{3}{4} \div \frac{9}{16} = \frac{3}{4} \times \frac{16}{9} = \frac{4}{3} = 1\frac{1}{3}$

3. $\frac{3}{16} \div \frac{7}{8} = \frac{3}{16} \times \frac{8}{7} = \frac{3}{14}$

5. $\frac{7}{9} \div \frac{2}{3} = \frac{7}{9} \times \frac{3}{2} = \frac{7}{6} = 1\frac{1}{6}$

7. $\frac{9}{16} \div \frac{21}{32} = \frac{9}{16} \times \frac{32}{21} = \frac{6}{7}$

9. $\frac{8}{25} \div \frac{16}{35} = \frac{8}{25} \times \frac{35}{16} = \frac{7}{10}$

11. $\frac{5}{36} \times \frac{54}{15} = \frac{1}{2}$

13. $\frac{7}{13} \times \frac{32}{11} = 1\frac{3}{11}$

15. $\frac{35}{64} \times \frac{32}{15} = 1\frac{1}{6}$

17. $\frac{92}{225} \times \frac{175}{23} = 3\frac{1}{9}$

19. $\frac{105}{121} \times \frac{363}{320} = \frac{63}{64}$

***Problems 6-5***

1. $\frac{1}{8} + \frac{3}{8} = \frac{4}{8} = \frac{1}{2}$

3. $\frac{2}{7} + \frac{3}{7} = \frac{5}{7}$

5. $\frac{11}{12} - \frac{5}{12} = \frac{6}{12} = \frac{1}{2}$

7. $\frac{9}{14} - \frac{5}{14} = \frac{4}{14} = \frac{2}{7}$

9. $\frac{5}{12} + \frac{7}{12} + \frac{11}{12} = \frac{23}{12} = 1\frac{11}{12}$

11. $\frac{3}{16} - \frac{5}{16} + \frac{7}{16} = \frac{5}{16}$

13. $\frac{7}{24} + \frac{11}{24} - \frac{5}{24} = \frac{13}{24}$

15. $\frac{7}{9} - \frac{5}{9} - \frac{4}{9} = -\frac{2}{9}$

17. $\frac{11}{16} + \frac{5}{16} - \frac{7}{16} = \frac{9}{16}$

19. $\frac{17}{18} - \frac{11}{18} + \frac{5}{18} = \frac{11}{18}$

***Problems 6-6***

1. $6 = 2 \cdot 3$
   $15 = 3 \cdot 5$
   LCM $= 2 \cdot 3 \cdot 5 = 30$

3. $55 = 5 \cdot 11$
   $65 = 5 \cdot 13$
   LCM $= 5 \cdot 11 \cdot 13 = 715$

5. $84 = 2 \cdot 2 \cdot 3 \cdot 7$
   $90 = 2 \cdot 3 \cdot 3 \cdot 5$
   LCM $= 2 \cdot 2 \cdot 3 \cdot 3 \cdot 5 \cdot 7 = 1260$

7. $22 = 2 \cdot 11$
   $26 = 2 \cdot 13$
   LCM $= 2 \cdot 11 \cdot 13 = 286$

9. $8 = 2 \cdot 2 \cdot 2$
   $24 = 2 \cdot 2 \cdot 2 \cdot 3$
   $36 = 2 \cdot 2 \cdot 3 \cdot 3$
   LCM $= 2 \cdot 2 \cdot 2 \cdot 3 \cdot 3 = 72$

11. 
$63 = 3\cdot3\cdot7$
$75 = 3\cdot5\cdot5$
$105 = 3\cdot5\cdot7$
$\text{LCM} = 3\cdot3\cdot5\cdot5\cdot7 = 1575$

13. 
$18 = 2\cdot3\cdot3$
$36 = 2\cdot2\cdot3\cdot3$
$72 = 2\cdot2\cdot2\cdot3\cdot3$
$\text{LCM} = 2\cdot2\cdot2\cdot3\cdot3 = 72$

15. 
$20 = 2\cdot2\cdot5$
$52 = 2\cdot2\cdot13$
$65 = 5\cdot13$
$\text{LCM} = 2\cdot2\cdot5\cdot13 = 260$

17. 
$50 = 2\cdot5\cdot5$
$70 = 2\cdot5\cdot7$
$175 = 5\cdot5\cdot7$
$\text{LCM} = 2\cdot5\cdot5\cdot7 = 350$

19. 
$45 = 3\cdot3\cdot5$
$75 = 3\cdot5\cdot5$
$225 = 3\cdot3\cdot5\cdot5$
$\text{LCM} = 3\cdot3\cdot5\cdot5 = 225$

***Problems 6-7***

1. $\frac{1}{4} + \frac{1}{6} = \frac{3}{12} + \frac{2}{12} = \frac{5}{12}$

3. $\frac{2}{5} + \frac{4}{15} = \frac{6}{15} + \frac{4}{15} = \frac{10}{15} = \frac{2}{3}$

5. $\frac{2}{3} - \frac{1}{6} = \frac{4}{6} - \frac{1}{6} = \frac{3}{6} = \frac{1}{2}$

7. $\frac{7}{8} - \frac{3}{16} = \frac{14}{16} - \frac{3}{16} = \frac{11}{16}$

9. $\frac{17}{24} - \frac{5}{6} = \frac{17}{24} - \frac{20}{24} = -\frac{3}{24} = -\frac{1}{8}$

11. $\frac{5}{12} - \frac{13}{16} = \frac{20}{48} - \frac{39}{48} = -\frac{19}{48}$

13. $\frac{2}{3} + \frac{1}{6} + \frac{1}{12} = \frac{8}{12} + \frac{2}{12} + \frac{1}{12} = \frac{11}{12}$

15. $\frac{4}{15} + \frac{3}{25} + \frac{1}{3} = \frac{20}{75} + \frac{9}{75} + \frac{25}{75} = \frac{54}{75} = \frac{18}{25}$

17. $\frac{11}{42} + \frac{2}{3} - \frac{5}{28} = \frac{22}{84} + \frac{56}{84} - \frac{15}{84} = \frac{63}{84} = \frac{3}{4}$

19. $\frac{9}{56} - \frac{13}{42} - \frac{1}{21} = \frac{27}{168} - \frac{52}{168} - \frac{8}{168} = -\frac{33}{168} = -\frac{11}{56}$

21. $\frac{4}{55} - \frac{3}{25} + \frac{3}{5} = \frac{20}{275} - \frac{33}{275} + \frac{165}{275} = \frac{152}{275}$

23. $\frac{3}{65} + \frac{2}{15} - \frac{4}{39} = \frac{9}{195} + \frac{26}{195} - \frac{20}{195} = \frac{15}{195} = \frac{1}{13}$

***Problems 6-8***

1. $\frac{17}{2} = 8\frac{1}{2}$

3. $\frac{37}{4} = 9\frac{1}{4}$

5. $\frac{19}{5} = 3\frac{4}{5}$

7. $\frac{43}{7} = 6\frac{1}{7}$

9. $\frac{27}{4} = 6\frac{3}{4}$

11. $\frac{67}{7} = 9\frac{4}{7}$

13. $\frac{87}{16} = 5\ \frac{7}{16}$

15. $\frac{89}{12} = 7\ \frac{5}{12}$

17. $\frac{60}{9} = 6\ \frac{6}{9} = 6\ \frac{2}{3}$

19. $\frac{37}{15} = 2\ \frac{7}{15}$

21. $10\ \frac{1}{2} = \frac{21}{2}$

23. $4\ \frac{3}{16} = \frac{67}{16}$

25. $3\ \frac{5}{8} = \frac{29}{8}$

27. $9\ \frac{1}{3} = \frac{28}{3}$

29. $5\ \frac{1}{6} = \frac{31}{6}$

31. $8\ \frac{7}{8} = \frac{71}{8}$

33. $12\ \frac{1}{3} = \frac{37}{3}$

35. $7\ \frac{1}{3} = \frac{22}{3}$

37. $10\ \frac{3}{5} = \frac{53}{5}$

39. $4\ \frac{3}{16} = \frac{67}{16}$

***Problems 6-9***

1. $\frac{3}{8} \times \frac{10}{3} = \frac{30}{24} = 1\ \frac{6}{24} = 1\ \frac{1}{4}$

3. $\frac{3}{4} \times \frac{8}{5} = \frac{24}{20} = 1\ \frac{4}{20} = 1\ \frac{1}{5}$

5. $\frac{3}{8} \times \frac{5}{3} = \frac{15}{24} = \frac{5}{8}$

7. $\frac{9}{4} \times 2\ \frac{1}{5} = \frac{9}{4} \times \frac{11}{5} = \frac{99}{20} = 4\ \frac{19}{20}$

9. $3 \times 3\ \frac{1}{4} = 3 \times \frac{13}{4} = \frac{39}{4} = 9\ \frac{3}{4}$

11. $2\ \frac{1}{3} \times 3\ \frac{1}{4} = \frac{7}{3} \times \frac{13}{4} = \frac{91}{12} = 7\ \frac{7}{12}$

13. $3\ \frac{1}{3} \times 4\ \frac{1}{5} = \frac{10}{3} \times \frac{21}{5} = \frac{210}{15} = 14$

15. $3\ \frac{3}{8} \times 6\ \frac{1}{3} = \frac{27}{8} \times \frac{19}{3} = \frac{513}{24} = 21\ \frac{9}{24} = 21\ \frac{3}{8}$

17. $2\ \frac{5}{6} \times 3\ \frac{2}{3} = \frac{17}{3} \times \frac{11}{2} = \frac{187}{18} = 10\ \frac{7}{18}$

19. $5\ \frac{5}{6} \times 4\ \frac{2}{5} = \frac{35}{6} \times \frac{22}{5} = \frac{770}{30} = 25\ \frac{20}{30} = 25\ \frac{2}{3}$

21. $\frac{7}{2} \div \frac{4}{3} = \frac{7}{2} \times \frac{3}{4} = \frac{21}{8} = 2\ \frac{5}{8}$

23. $\frac{7}{16} \div \frac{1}{4} = \frac{7}{16} \times 4 = \frac{28}{16} = 1\ \frac{12}{16} = 1\ \frac{3}{4}$

25. $3 \div \frac{7}{5} = 3 \times \frac{5}{7} = \frac{15}{7} = 2\ \frac{1}{7}$

27. $6 \div \frac{5}{8} = 6 \times \frac{8}{5} = \frac{48}{5} = 9\ \frac{3}{5}$

29. $\frac{9}{4} \div 1\ \frac{2}{3} = \frac{9}{4} \div \frac{5}{3} = \frac{9}{4} \times \frac{3}{5} = \frac{27}{20} = 1\ \frac{7}{20}$

31. $\frac{7}{4} \div 2\ \frac{4}{5} = \frac{7}{4} \div \frac{14}{5} = \frac{7}{4} \times \frac{5}{14} = \frac{35}{56} = \frac{5}{8}$

33. $3\ \frac{1}{2} \div 4\ \frac{1}{3} = \frac{7}{2} \div \frac{13}{3} = \frac{7}{2} \times \frac{3}{13} = \frac{21}{26}$

35. $2\ \frac{1}{3} \div 6\ \frac{1}{2} = \frac{7}{3} \div \frac{13}{2} = \frac{7}{3} \times \frac{2}{13} = \frac{14}{39}$

37. $5\ \frac{3}{5} \div 3\ \frac{1}{3} = \frac{28}{5} \div \frac{10}{3} = \frac{28}{5} \times \frac{3}{10} = \frac{84}{50} = 1\ \frac{34}{50} = 1\ \frac{17}{25}$

39. $4\ \frac{2}{5} \div 1\ \frac{8}{25} = \frac{22}{5} \div \frac{33}{25} = \frac{22}{5} \times \frac{25}{33} = \frac{550}{165} = 3\ \frac{55}{165} = 3\ \frac{1}{3}$

***Problems 6-10***

1. $\frac{16}{5} + \frac{10}{3} = \frac{48}{15} + \frac{50}{15} = \frac{98}{15} = 6\ \frac{8}{15}$

3. $4\ \frac{5}{6} + \frac{5}{7} = \frac{29}{6} + \frac{5}{7} = \frac{203}{42} + \frac{30}{42} = \frac{233}{42} = 5\ \frac{23}{42}$

5. $2\ \frac{1}{3} + \frac{9}{5} = \frac{7}{3} + \frac{9}{5} = \frac{35}{15} + \frac{27}{15} = \frac{62}{15} = 4\ \frac{2}{15}$

7. $3\ \frac{7}{8} + 4\ \frac{1}{2} = \frac{31}{8} + \frac{9}{2} = \frac{31}{8} + \frac{36}{8} = \frac{67}{8} = 8\ \frac{3}{8}$

9. $\frac{7}{8} - \frac{7}{32} = \frac{28}{32} - \frac{7}{32} = \frac{21}{32}$

11. $\frac{10}{3} - \frac{1}{4} = \frac{40}{12} - \frac{3}{12} = \frac{37}{12} = 3\ \frac{1}{12}$

13. $3\ \frac{5}{6} - \frac{9}{4} = \frac{23}{6} - \frac{9}{4} = \frac{46}{12} - \frac{27}{12} = \frac{19}{12} = 1\ \frac{7}{12}$

15. $3\ \frac{1}{8} - 1\ \frac{1}{4} = \frac{25}{8} - \frac{5}{4} = \frac{25}{8} - \frac{10}{8} = \frac{15}{8} = 1\ \frac{7}{8}$

17. $4\ \frac{5}{6} - 2\ \frac{2}{3} = \frac{29}{6} - \frac{8}{3} = \frac{29}{6} - \frac{16}{6} = \frac{13}{6} = 2\ \frac{1}{6}$

19. $\frac{9}{4} + \frac{7}{2} - 2\ \frac{1}{3} = \frac{9}{4} + \frac{7}{2} - \frac{7}{3} = \frac{27}{12} + \frac{42}{12} - \frac{28}{12} = \frac{41}{12} = 3\ \frac{5}{12}$

21. $7\ \frac{1}{2} - 2\ \frac{1}{3} + 3\ \frac{1}{4} = \frac{15}{2} - \frac{7}{3} + \frac{13}{4} = \frac{90}{12} - \frac{28}{12} + \frac{39}{12} = \frac{101}{12} = 8\ \frac{5}{12}$

23. $2\ \frac{7}{8} + 4\ \frac{3}{4} - 3\ \frac{11}{16} = \frac{23}{8} + \frac{19}{4} - \frac{59}{16} = \frac{46}{16} + \frac{76}{16} - \frac{59}{16} = \frac{63}{16} = 3\ \frac{15}{16}$

25. $3\ \frac{7}{16} - \frac{15}{32} + 2\ \frac{3}{8} = \frac{55}{16} - \frac{15}{32} + \frac{19}{8} = \frac{110}{32} - \frac{15}{32} + \frac{76}{32} = \frac{171}{32} = 5\ \frac{11}{32}$

27. $7\ \frac{2}{3} - 3\ \frac{2}{5} + 2\ \frac{4}{15} = \frac{23}{3} - \frac{17}{5} + \frac{34}{15} = \frac{115}{15} - \frac{51}{15} + \frac{34}{15} = \frac{98}{15} = 6\ \frac{8}{15}$

29. $8\ \frac{3}{8} + 4\ \frac{2}{3} - 3\ \frac{1}{12} = \frac{67}{8} + \frac{14}{3} - \frac{37}{12} = \frac{201}{24} + \frac{112}{24} - \frac{74}{24} = \frac{239}{24} = 9\ \frac{23}{24}$

***Problems 6-11***

1. $\frac{3}{8} + \frac{3}{32} = \frac{12}{32} + \frac{3}{32} = \frac{15}{32} = 0.469$

   $\frac{3}{8} - \frac{3}{32} = \frac{12}{32} - \frac{3}{32} = \frac{9}{32} = 0.281$

   $\frac{3}{8} \times \frac{3}{32} = \frac{9}{256} = 0.0352$

   $\frac{3}{8} \div \frac{3}{32} = \frac{3}{8} \times \frac{32}{3} = 4.00$

3. $\frac{7}{16} + \frac{2}{7} = \frac{49}{112} + \frac{32}{112} = \frac{81}{112} = 0.723$

   $\frac{7}{16} - \frac{2}{7} = \frac{49}{112} - \frac{32}{112} = \frac{17}{112} = 0.152$

   $\frac{7}{16} \times \frac{2}{7} = \frac{14}{112} = 0.125$

   $\frac{7}{16} \div \frac{2}{7} = \frac{7}{16} \times \frac{7}{2} = \frac{49}{32} = 1.53$

5. $\frac{5}{8} + \frac{37}{64} = \frac{40}{64} + \frac{37}{64} = \frac{77}{64} = 1.20$

   $\frac{5}{8} - \frac{37}{64} = \frac{40}{64} - \frac{37}{64} = \frac{3}{64} = 0.0469$

   $\frac{5}{8} \times \frac{37}{64} = \frac{185}{512} = 0.361$

   $\frac{5}{8} \div \frac{37}{64} = \frac{5}{8} \times \frac{64}{37} = \frac{320}{296} = 1.08$

7. $\frac{3}{5} + \frac{3}{10} = \frac{6}{10} + \frac{3}{10} = \frac{9}{10} = 0.900$

   $\frac{3}{5} - \frac{3}{10} = \frac{6}{10} - \frac{3}{10} = \frac{3}{10} = 0.300$

   $\frac{3}{5} \times \frac{3}{10} = \frac{9}{50} = 0.180$

   $\frac{3}{5} \div \frac{3}{10} = \frac{3}{5} \times \frac{10}{3} = \frac{30}{15} = 2.00$

9. $3\ \frac{3}{8} + 2\ \frac{1}{3} = \frac{27}{8} + \frac{7}{3} = \frac{81}{24} + \frac{56}{24} = \frac{137}{24} = 5.71$

$3\ \frac{3}{8} - 2\ \frac{1}{3} = \frac{27}{8} - \frac{7}{3} = \frac{81}{24} - \frac{56}{24} = \frac{25}{24} = 1.04$

$3\ \frac{3}{8} \times 2\ \frac{1}{3} = \frac{27}{8} \times \frac{7}{3} = \frac{189}{24} = 7.88$

$3\ \frac{3}{8} \div 2\ \frac{1}{3} = \frac{27}{8} \times \frac{3}{7} = \frac{81}{56} = 1.45$

11. $4\ \frac{2}{5} + 2\ \frac{15}{64} = \frac{22}{5} + \frac{143}{64} = \frac{1408}{320} + \frac{715}{320} = \frac{2123}{320} = 6.63$

$4\ \frac{2}{5} - 2\ \frac{15}{64} = \frac{1408}{320} - \frac{715}{320} = \frac{693}{320} = 2.17$

$4\ \frac{2}{5} \times 2\ \frac{15}{64} = \frac{22}{5} \times \frac{143}{64} = \frac{3146}{320} = 9.83$

$4\ \frac{2}{5} \div 2\ \frac{15}{64} = \frac{22}{5} \times \frac{64}{143} = \frac{1408}{715} = 1.97$

13. $7\ \frac{1}{2} + 3\ \frac{5}{9} = \frac{15}{2} + \frac{32}{9} = \frac{135}{18} + \frac{64}{18} = \frac{199}{18} = 11.1$

$7\ \frac{1}{2} - 3\ \frac{5}{9} = \frac{135}{18} - \frac{64}{18} = \frac{71}{18} = 3.94$

$7\ \frac{1}{2} \times 3\ \frac{5}{9} = \frac{15}{2} \times \frac{32}{9} = \frac{480}{18} = 26.7$

$7\ \frac{1}{2} \div 3\ \frac{5}{9} = \frac{15}{2} \times \frac{9}{32} = \frac{135}{64} = 2.11$

15. $4\ \frac{7}{8} + 8\ \frac{7}{16} = \frac{39}{8} + \frac{135}{16} = \frac{78}{16} + \frac{135}{16} = \frac{213}{16} = 13.3$

$4\ \frac{7}{8} - 8\ \frac{7}{16} = \frac{78}{16} - \frac{135}{16} = -\frac{57}{16} = -3.56$

$4\ \frac{7}{8} \times 8\ \frac{7}{16} = \frac{39}{8} \times \frac{135}{16} = \frac{5265}{128} = 41.1$

$4\ \frac{7}{8} \div 8\ \frac{7}{16} = \frac{39}{8} \times \frac{16}{135} = \frac{624}{1080} = 0.578$

17. $3\ \frac{2}{3} + 6\ \frac{2}{3} = \frac{11}{3} + \frac{20}{3} = \frac{31}{3} = 10.3$

$3\ \frac{2}{3} - 6\ \frac{2}{3} = \frac{11}{3} - \frac{20}{3} = -\frac{9}{3} = -3.00$

$3\ \frac{2}{3} \times 6\ \frac{2}{3} = \frac{11}{3} \times \frac{20}{3} = \frac{220}{9} = 24.4$

$3\ \frac{2}{3} \div 6\ \frac{2}{3} = \frac{11}{3} \times \frac{3}{20} = \frac{33}{60} = 0.550$

19. $1\ \frac{7}{12} + 3\ \frac{2}{3} = \frac{19}{12} + \frac{11}{3} = \frac{19}{12} + \frac{44}{12} = \frac{63}{12} = 5.25$

$1\ \frac{7}{12} - 3\ \frac{2}{3} = \frac{19}{12} + \frac{44}{12} = -\frac{25}{12} = -2.08$

$1\ \frac{7}{12} \times 3\ \frac{2}{3} = \frac{19}{12} \times \frac{11}{3} = \frac{209}{36} = 5.81$

$1\ \frac{7}{12} \div 3\ \frac{2}{3} = \frac{19}{12} \times \frac{3}{11} = \frac{57}{132} = 0.432$

# CHAPTER 7

***Problems 7-1***

1. $2 \cdot 2 \cdot 2 \cdot a \cdot a \cdot b \cdot b$
3. $2 \cdot 2 \cdot 2 \cdot 3 \cdot x \cdot x \cdot y$
5. $2 \cdot 3 \cdot 7 \cdot c \cdot c \cdot c$
7. $2 \cdot 2 \cdot 2 \cdot 11 \cdot y \cdot y \cdot z$
9. $3 \cdot 3 \cdot 3 \cdot 5 \cdot a \cdot a \cdot c \cdot c$

***Problems 7-2***

1. $a^3b^3$
3. $a^3b^3c^3$
5. $x^2y^3z^4$
7. $a^3b^3$
9. $a^4b^2c^3$
11. $a^4b^3c^2$
13. $78x^2y^3$
15. $32x^3y^2z^2$
17. $150a^2b^2c$
19. $270x^2y^4z^3$

***Problems 7-3***

1. $21a^4b^3$
3. $20a^{-4}b^{-5}$
5. $20x^6y^3z^4$
7. $6a^3b^{-2}c^{-1}$
9. $60x^5y^{-4}z^4$

***Problems 7-4***

1. $9a^2$
3. $8ab^2$
5. $5a^3bc^{-1}$
7. $\dfrac{xy}{5}$
9. $11x^{-5}y^{-1}z^{-4}$

***Problems 7-5***

1. $$\frac{4b}{15} \times \frac{3b}{8} = \frac{12b^2}{120} = \frac{b^2}{10}$$

3. $$\frac{3}{2b} \times \frac{5}{7b} = \frac{15}{14b^2} = \frac{15b^{-2}}{14}$$

5. $$\frac{5a}{9} \times \frac{6a^2}{7} = \frac{30a^3}{63} = \frac{2 \cdot 3 \cdot 5a^3}{3 \cdot 3 \cdot 7} = \frac{10a^3}{21}$$

7. $$\frac{3x^2}{10y} \times \frac{5x}{8y} = \frac{15x^3}{80y^2} = \frac{3 \cdot 5x^3}{2 \cdot 2 \cdot 2 \cdot 2 \cdot 5y^2} = \frac{3x^3y^{-2}}{16}$$

9. $$\frac{3}{4y^2} \times \frac{8xy}{9} = \frac{24xy}{36y^2} = \frac{2 \cdot 2 \cdot 2 \cdot 3xy}{2 \cdot 2 \cdot 3 \cdot 3y^2} = \frac{2xy^{-1}}{3}$$

11. $\frac{8x^{-1}y}{5} \times \frac{10x^2y^{-3}}{3z^{-1}} = \frac{80xy^{-2}}{15z^{-1}} = \frac{2\cdot2\cdot2\cdot2\cdot5xy^{-2}}{3\cdot5x} = \frac{16xy^{-2}z}{3}$

13. $\frac{11a^2}{32b^3} \times \frac{4a^2b}{33c} = \frac{44a^4b}{1060b^3c} = \frac{2\cdot2\cdot11a^4b}{2\cdot2\cdot2\cdot2\cdot2\cdot3\cdot11b^3c} = \frac{a^4b^{-2}c^{-1}}{24}$

15. $\frac{7x^2}{24y^2} \times \frac{4xy^{-2}}{21x} = \frac{28x^3y^{-2}}{504xy^2} = \frac{2\cdot2\cdot7x^3y^{-2}}{2\cdot2\cdot2\cdot3\cdot3\cdot7xy^2} = \frac{x^2y^{-4}}{18}$

17. $\frac{4a^3c^{-2}}{5b^2} \times \frac{7b^{-1}c^3}{16a^2} \times \frac{20a^3b^{-1}}{21c} = \frac{560a^6b^{-2}c}{1680a^2b^2c} = \frac{2\cdot2\cdot2\cdot2\cdot5\cdot7a^6b^{-2}c}{2\cdot2\cdot2\cdot2\cdot3\cdot5\cdot7a^2b^2c} = \frac{a^4b^{-4}}{3}$

19. $\frac{6xy^{-2}}{15z^2} \times \frac{25x^{-3}z^2}{18y} \times \frac{12y^2}{5x^{-1}z^{-2}} = \frac{1800x^{-2}z^2}{1350x^{-1}y} = \frac{2\cdot2\cdot2\cdot3\cdot3\cdot5\cdot5x^{-2}z^2}{2\cdot3\cdot3\cdot3\cdot5\cdot5x^{-1}y} = \frac{4x^{-1}y^{-1}z^2}{3}$

***Problems 7-6***

1. $\frac{3x}{8} \div \frac{7x^3}{3} = \frac{3x}{8} \times \frac{3}{7x^3} = \frac{9x}{56x^3} = \frac{9x^{-2}}{56}$

3. $\frac{7x}{16} \div \frac{21x^2}{32} = \frac{7x}{16} \times \frac{32}{21x^2} = \frac{224x}{336x^2} = \frac{2\cdot2\cdot2\cdot2\cdot2\cdot7\cdot x}{2\cdot2\cdot2\cdot2\cdot3\cdot7\cdot x^2} = \frac{2x^{-1}}{3}$

5. $\frac{16b}{3} \div \frac{8b^4}{27} = \frac{16b}{3} \times \frac{27}{8b^4} = \frac{432b}{24b^4} = \frac{2\cdot2\cdot2\cdot2\cdot3\cdot3\cdot3b}{2\cdot2\cdot2\cdot3b^4} = 18b^{-3}$

7. $\frac{14ab^2}{9c^3} \div \frac{35b}{c^2} = \frac{14ab^2}{9c^3} \times \frac{c^2}{35b} = \frac{14ab^2c^2}{315bc^3} = \frac{2\cdot7ab^2c^2}{3\cdot3\cdot5\cdot7bc^3} = \frac{2abc^{-1}}{45}$

9. $\frac{3xy^3}{10x^{-1}} \div \frac{3xy^3}{10x^{-1}} = \frac{3xy^3}{10x^{-1}} \times \frac{10x^{-1}}{3xy^3} = \frac{30y^3}{30y^3} = 1$

11. $\frac{32a}{45b^2} \div \frac{16a^2}{25bc} = \frac{32a}{45b^2} \times \frac{25bc}{16a^2} = \frac{800abc}{720a^2b^2} = \frac{2\cdot2\cdot2\cdot2\cdot2\cdot5\cdot5abc}{2\cdot2\cdot2\cdot2\cdot3\cdot3\cdot5a^2b^2} = \frac{10a^{-1}b^{-1}c}{9}$

13. $\frac{15a^4b}{64c^3} \div \frac{75ab}{32c^2} = \frac{15a^4b}{64c^3} \times \frac{32c^2}{75ab} = \frac{480a^4bc^2}{4800abc^3} = \frac{a^3c^{-1}}{10}$

15. $\frac{20x^3y^2}{21z^2} \div \frac{30xy}{7z^2} = \frac{20x^3y^2}{21z^2} \times \frac{7z^2}{30xy} = \frac{140x^3y^2z^2}{630xyz^2} = \frac{2\cdot2\cdot5\cdot7x^3y^2z^2}{2\cdot3\cdot3\cdot5\cdot7xyz^2} = \frac{2x^2y}{9}$

17. $\frac{21a^3b^{-2}}{4c^2} \div \frac{14a^{-1}b^2}{3c^{-4}} = \frac{21a^3b^{-2}}{4c^2} \times \frac{3c^{-4}}{14a^{-1}b^2} = \frac{63a^3b^{-2}c^{-4}}{56a^{-1}b^2c^2} = \frac{3\cdot3\cdot7a^3b^{-2}c^{-4}}{2\cdot2\cdot2\cdot7a^{-1}b^2c^2}$

$= \frac{9a^4b^{-4}c^{-6}}{8}$

19. $\frac{42yz^{-2}}{55x} \div \frac{21x^2y^2}{22z} = \frac{42yz^{-2}}{55x} \times \frac{22z}{21x^2y^2} = \frac{924yz^{-1}}{1155x^3y^2} = \frac{2\cdot2\cdot3\cdot7\cdot11yz^{-1}}{3\cdot5\cdot7\cdot aax^3y^2} = \frac{4x^{-3}y^{-1}z^{-1}}{5}$

***Problems 7-7***

1. $\frac{5x^{-1}}{7}$

3. $\frac{4x^{-1}}{7}$

5. $\frac{9}{12ab} = \frac{3a^{-1}b^{-1}}{4}$

7. $\frac{7}{14a} = \frac{a^{-1}}{2}$

9. $\frac{x}{2}$

11. $\frac{2x}{5}$

13. $\frac{15ab}{25} = \frac{3ab}{5}$

15. $\frac{15ab}{25} = \frac{3ab}{5}$

17. $\frac{10x}{9}$

19. $\frac{-2x}{7} = -\frac{2x}{7}$

21. $\frac{-5x}{24} = -\frac{5x}{24}$

23. $\frac{2ab+3b}{8}$

25. $\frac{7abc-3ac}{4d}$

27. $-\frac{2abc+3ab}{4x}$

29. $\frac{6xy-4xz+3yz}{5}$

***Problems 7-8***

1. $\frac{9x}{15} + \frac{2x}{15} = \frac{11x}{15}$

3. $\frac{4a}{6} + \frac{9a}{6} = \frac{13a}{6}$

5. $\frac{3x}{6} - \frac{4x}{6} = -\frac{x}{6}$

7. $\frac{8x}{10} - \frac{3x}{10} = \frac{5x}{10} = \frac{x}{2}$

9. $\frac{y}{2x} + \frac{5y}{8} = \frac{4y}{8x} + \frac{5xy}{8x} = \frac{4y+5xy}{8x}$

11. $\frac{3x}{4y} + \frac{2y}{3x} = \frac{9x^2}{12xy} + \frac{8y^2}{12xy} = \frac{9x^2+8y^2}{12xy}$

13. $\frac{2b}{5x} + \frac{3}{10y} = \frac{4by}{10xy} + \frac{3x}{10xy} = \frac{4by+3x}{10xy}$

15. $\frac{5ab^2}{16c} - \frac{15ab}{32c^2} = \frac{10ab^2c}{32c^2} - \frac{15ab}{32c^2} = \frac{10ab^2c-15ab}{32c^2}$

17. $\frac{3b}{4} + \frac{3b}{8} - \frac{15b}{16} = \frac{12b}{16} + \frac{6b}{16} - \frac{15b}{16} = \frac{3b}{16}$

19. $\frac{2a}{7} + \frac{7a}{15} - \frac{a}{3} = \frac{30a}{105} + \frac{49a}{105} - \frac{35a}{105} = \frac{44a}{105}$

21. $\frac{x}{10y^2} - \frac{2x}{5y^2} + \frac{x}{3y^2} = \frac{3x}{30y^2} - \frac{12x}{30y^2} + \frac{10x}{30y^2} = \frac{x}{30y^2}$

23. $\frac{4x}{3yz} + \frac{2y}{5x} - \frac{z}{6y} = \frac{40x^2}{30xyz} + \frac{12y^2z}{30xyz} - \frac{5xz^2}{30xyz} = \frac{40x^2+12y^2z-5xz^2}{30xyz}$

25. $\frac{4a}{33b} + \frac{5b}{55c} - \frac{3c}{11a} = \frac{40a^2c}{330abc} + \frac{30ab^2}{330abc} - \frac{90bc^2}{330abc} = \frac{40a^2c+30ab^2-90bc^2}{330abc}$

$= \frac{4a^2c+3ab^2-9bc^2}{33abc}$

27. $\frac{5y^2}{6x^2y} + \frac{4xy}{15y^2} - \frac{2x^2}{9y} = \frac{75y^3}{90x^2y^2} + \frac{24x^3y}{90x^2y^2} - \frac{20x^4y}{90x^2y^2} = \frac{75y^3+24x^3y-20x^4y}{90x^2y^2}$

29. $\frac{3b}{20a^2} + \frac{7c}{10ab} - \frac{5a}{12b^3} = \frac{9b^4}{60a^2b^3} + \frac{42ab^2c}{60a^2b^3} - \frac{25a^3}{60a^2b^3} = \frac{9b^4+42ab^2c-25a^3}{60a^2b^3}$

# CHAPTER 8

***Problems 8-1***

1. $a - 7 = 3$, $a = 3 + 7$, $a = 10$

3. $x = 2$

5. $x = 3$

7. $x = 9$

9. $b = 4$

11. $x = -4$

13. $5y = 20$
    $y = 4$

15. $b = -9$

17. $\frac{a}{4} = 3$

    $a = 12$

19. $x = -7$

21. $c = 12$

23. $a = -10$

25. $x = 16$

27. $x = 12$

29. $\frac{3}{5y} = 6$, $\frac{5y}{3} = \frac{1}{6}$

    $y = \frac{3}{5 \cdot 6} = \frac{1}{10}$

31. $\frac{3x}{4} = 6$, $x = \frac{6 \cdot 4}{3}$

    $x = 8$

33. $\frac{7}{2y} = 14$, $\frac{2y}{7} = \frac{1}{14}$

    $y = \frac{7}{2 \cdot 14} = \frac{1}{4}$

35. $\frac{5y}{-3} = 10$

    $y = \frac{-3 \cdot 10}{5} = -6$

37. $\frac{-6}{7y} = 3$, $\frac{7y}{-6} = \frac{1}{3}$

    $a = \frac{-6}{7 \cdot 3} = -\frac{2}{7}$

39. $\frac{-8}{3b} = 2$, $\frac{3b}{-8} = \frac{1}{2}$

    $b = \frac{-8}{3 \cdot 2} = -1\frac{1}{3}$

41. $\frac{x}{4} = \frac{3}{2} = \frac{3 \cdot 4}{2} = 6$

43. $a = \frac{3 \cdot 5}{2 \cdot 7} = \frac{15}{14} = 1\frac{1}{14}$

45. $\frac{3 \cdot 5}{10 \cdot 14} = x = \frac{3}{28}$

47. $\frac{2a}{7} = \frac{4}{21}$ $a = \frac{7 \cdot 4}{2 \cdot 21} = \frac{2}{3}$

49. $\frac{5x}{3} = \frac{9}{4}$ $x = \frac{9 \cdot 3}{5 \cdot 4} = 1\frac{7}{20}$

***Problems 8-2***

1. $x^2 = 25$
   $x = 5$

3. $\frac{a^2}{4} = 18$
   $a^2 = 54$
   $a = 7.35$

5. $\frac{a^2}{4} = 14,\ a^2 = 56$
   $a = 7.48$

7. $\frac{6}{b^2} = 4,\ \frac{b^2}{6} = \frac{1}{4},\ b^2 = \frac{6}{4}$
   $b = 1.22$

9. $12 = \frac{72}{a^2},\ a^2 = \frac{72}{12}$
e
   $a = 2.45$

11. $\sqrt{a} = 7$
    $a = 49$

13. $\frac{\sqrt{x}}{3} = 2,\ \sqrt{x} = 6$
    $x = 36$

15. $\frac{\sqrt{a}}{b} = 2.3,\ \sqrt{a} = 13.8$
    $a = 190$

17. $2 = \frac{1}{2\sqrt{x}},\ \sqrt{x} = \frac{1}{4},\ x = \frac{1}{16}$

19. $3 = \frac{1}{4\sqrt{x}},\ \sqrt{x} = \frac{1}{12},\ x = \frac{1}{144}$

21. $\frac{3}{4\sqrt{a}} = 6,\ \frac{4\sqrt{a}}{3} = \frac{1}{6},\ \sqrt{a} = \frac{3}{24},\ a = \frac{1}{64}$

23. $\frac{2\sqrt{x}}{3} = 6,\ \sqrt{x} = \frac{18}{2},\ x = 81$

25. $2ax = 12,\ ax = \frac{12}{2},\ ax = 6$
    $a = \frac{6}{x},\ x = \frac{6}{a}$

27. $8bc = 14,\ bc = \frac{14}{8},\ bc = \frac{7}{4}$
    $b = \frac{7}{4c},\ c = \frac{7}{4b}$

29. $P = \frac{E^2}{R},\ R = \frac{E^2}{P},\ \frac{E^2}{R} = P,\ E^2 = PR,\ \sqrt{E^2} = \sqrt{PR},\ E = \sqrt{PR}$

31. $B_C = 2\pi fC;\ f = \frac{B_C}{2\pi C};\ C = \frac{B_C}{2\pi f}$

33. $3 = \frac{1}{2a\sqrt{b}},\ a = \frac{1}{6\sqrt{b}},\ \sqrt{b} = \frac{1}{6a},\ (\sqrt{b})^2 = (\frac{1}{6a})^2,\ b = \frac{1}{36a^2}$

35. $5 = \frac{6}{\sqrt{xy}},\ \sqrt{xy} = \frac{6}{5},\ (\sqrt{xy})^2 = (\frac{6}{5})^2,\ xy = \frac{36}{25},\ x = \frac{36}{25y};\ y = \frac{36}{25x}$

37. $X_L = 2\pi fL,\ 2\pi fL = X_L,\ f = \frac{X_L}{2\pi L},\ L = \frac{X_L}{2\pi f}$

39. $fr = \frac{1}{2\pi\sqrt{LC}},\ \sqrt{LC} = \frac{1}{2\pi fr},\ (\sqrt{LC})^2 = (\frac{1}{2\pi fr})^2,\ LC = \frac{1}{(2\pi f)^2} = L = \frac{1}{(2\pi f_r)^2 C};$
    $C = \frac{1}{(2\pi f_r)^2 L}$

***Problems 8-3***

1. $R_1 = R_T - R_2 = 12.4\ k\Omega - 5.6\ k\Omega = 6.8\ k\Omega$

3. $G_1 = G_T - G_2 = 54.9\ \mu S - 37\ \mu S = 17.9\ \mu S$

5. $A_V = \frac{V_o}{V_{in}} \rightarrow V_o = A_V\ V_{in} = 150 \cdot 55\ mV = 8.25\ V$

7. $R = \frac{E^2}{P} = \frac{(15\ V)^2}{2W} = 113\ \Omega$

9. $R = \frac{E^2}{P} = \frac{(450\ mV)^2}{150\ mW} = 1.35\ \Omega$

11. $R = \frac{E^2}{P} = \frac{(32\ V)^2}{30\ mW} = 34.1\ k\Omega$

13. $E^2 = PR,\ \sqrt{E^2} = \sqrt{PR},\ E = \sqrt{PR},\ E = (6.73\ W \cdot 27\ \Omega)^{1/2} = 13.5\ V$

15. $E^2 = PR,\ \sqrt{E^2} = \sqrt{PR},\ E = \sqrt{PR},\ E = (10.3\ W \cdot 1.8\ k\Omega)^{1/2} = 136\ V$

17. $E^2 = PR,\ \sqrt{E^2} = \sqrt{PR},\ E = \sqrt{PR},\ E = (1.73\ W \cdot 47\ k\Omega)^{1/2} = 285\ V$

19. $I = (\frac{P}{R})^{1/2},\ I = (\frac{600\ mW}{680\ \Omega})^{1/2} = 29.7\ mA$

21. $I = (\frac{P}{R})^{1/2},\ I = (\frac{500\ mW}{2.7\ k\Omega})^{1/2} = 13.6\ mA$

23. $I = (\frac{P}{R})^{1/2},\ I = (\frac{14.7\ W}{100\ \Omega})^{1/2} = 383\ mA$

25. $R = \frac{P}{I^2},\ R = \frac{500\ mW}{(10\ mA)^2} = 5.00\ k\Omega$

27. $R = \frac{P}{I^2},\ R = \frac{600\ mW}{(47.3\ mA)^2} = 268\ \Omega$

29. $R = \frac{P}{I^2},\ R = \frac{104\ W}{(3.77\ A)^2} = 7.32\ \Omega$

31. $N_p = N_s\ (\frac{Z_p}{Z_s})^{1/2} = 100\ (\frac{2.5\ k\Omega}{8\ \Omega})^{1/2} = 1770$

33. $N_p = N_s\ (\frac{Z_p}{Z_s})^{1/2} = 50\ (\frac{40\ \Omega}{500\ \Omega})^{1/2} = 14.1$

35. $N_p = N_s\ (\frac{Z_p}{Z_s})^{1/2} = 25\ (\frac{750\ \Omega}{75\ \Omega})^{1/2} = 79.1$

37. $Z_s = \frac{Z_p N_s^2}{N_p^2} = \frac{1\ k\Omega \cdot (100)^2}{(1200)^2} = 6.94\ \Omega$

39. $Z_s = \frac{Z_p N_s^2}{N_p^2} = \frac{2.3\ \Omega \cdot (1800)^2}{(100)^2} = 745\ \Omega$

41. $Z_s = \frac{Z_p N_s^2}{N_p^2} = \frac{325\ \Omega \cdot (60)^2}{(800)^2} = 1.83\ \Omega$

43. $N_s = N_p\ (\frac{Z_s}{Z_p})^{1/2} = 3000\ (\frac{12\ \Omega}{1.5\ k\Omega})^{1/2} = 268$

45. $N_s = N_p\ (\frac{Z_s}{Z_p})^{1/2} = 2000\ (\frac{4\ \Omega}{800\ \Omega})^{1/2} = 141$

47. $Z_p = \frac{Z_s N_p^2}{N_s^2} = \frac{2\ \text{k}\Omega \cdot (100)^2}{(2500)^2} = 3.20\ \Omega$

49. $Z_p = \frac{Z_s N_p^2}{N_s^2} = \frac{40\ \Omega \cdot (5000)^2}{(1000)^2} = 1.00\ \text{k}\Omega$

51. $C = \frac{1}{2\pi f X_C} = \frac{1}{2\pi \cdot 2.7\ \text{kHZ} \cdot 700\ \Omega} = 84.2\ \text{nF}$

53. $f = \frac{1}{2\pi C X_C} = \frac{1}{2\pi \cdot 10\ \text{nF} \cdot 5\ \text{k}\Omega} = 3.18\ \text{kHz}$

55. $L = \frac{X_L}{2\pi f} = \frac{3.85\ \text{k}\Omega}{2\pi \cdot 4.8\ \text{kHz}} = 128\ \text{mH}$

57. $f = \frac{X_L}{2\pi L} = \frac{170\ \Omega}{2\pi \cdot 54.1\ \text{mH}} = 500\ \text{Hz}$

59. $L = \frac{1}{C(2\pi f_r)^2} = \frac{1}{100\ \text{pF}(2\pi \cdot 30\ \text{kHz})^2} = 281\ \text{mH}$

61. $C = \frac{1}{L(2\pi f_r)^2} = \frac{1}{100\ \text{mH}(2\pi \cdot 100\ \text{Hz})^2} = 25.3\ \mu\text{F}$

# CHAPTER 9

***Problems 9-1***

1. $3(a + 3) = 3a + 9$
3. $4(a - 1) = 4a - 4$
5. $7(4 - b) = 28 - 7b$
7. $3a - 2(3a - 1) = 3a - 6a + 2 = -3a + 2$
9. $6a - 3(4a + 2) = 6a - 12a - 6 = -6a - 6$
11. $4(2x - 3) - 2x = 8x - 12 - 2x = 6x - 12$
13. $3x(2 + 2y) = 6x + 6xy$
15. $3a(2a + 3b) = 6a^2 + 9ab$
17. $3(3x^2 + 6x - 12) = 9x^2 + 18x - 36$
19. $2(4a^2 - 3a + 2) = 8a^2 - 6a + 4$
21. $4a(a + b) - 3a(a + 5b) = 4a^2 + 4ab - 3a^2 - 15ab = a^2 - 11ab$
23. $a(2ab + 2b) - 3(a^2b - 2ab) = 2a^2b + 2ab - 3a^2b + 6ab = -a^2b + 8ab$
25. $2x(4xy - 3y) - 3(x^2y + xy) = 8x^2y - 6xy - 3x^2y - 3xy = 5x^2y - 9xy$
27. $2x(3xy - 4y) - 4(x^2y + 3xy) = 6x^2y - 8xy - 4x^2y - 12xy = 2x^2y - 20xy$
29. $x(x^2 + 8x + 15) + 2x(x^2 - x - 6) = x^3 + 8x^2 + 15x + 2x^3 - 2x^2 - 12x$
    $= 3x^3 + 6x^2 + 3x$
31. $4a(a^2 - 3a + 6) - 3a(a^2 + 5a - 6) = 4a^3 - 12a^2 + 24a - 3a^3 - 15a^2 + 18a$
    $= a^3 - 27a^2 + 42a$
33. $3a^2 + 2a(a - b + 2) - 2a(2a + b) = 3a^2 + 2a^2 - 2ab + 4a - 4a^2 - 2ab$
    $= a^2 - 4ab + 4a$
35. $4x(2x - 3y + 2) - 2x^2 - 3x(x + 2y - 3)$
    $= 8x^2 - 12xy + 8x - 2x^2 - 3x^2 - 6xy + 9x = 3x^2 - 18xy + 17x$

***Problems 9-2***

1. $\frac{44x^2}{3}$
3. $5y^2$
5. $8a^2 + 9a$
7. $5b^2 - 6b$
9. $6a^2 - 16a + 8$
11. $2a^2 - 6a + \frac{9}{2}$
13. $-2x^2 + 7x$
15. $8x^2 - 15x$
17. $5b^2 - 6b$
19. $\frac{6x^2}{5} + \frac{21b}{4}$

***Problems 9-3***

1. $a^2 + 6a + 9$
3. $a^2 - 12a + 36$
5. $4x^2 - 16x + 16$
7. $16x^2 + 24x + 9$
9. $4a^2 + 12ab + 9b^2$
11. $x^6 - 8x^3y + 16y^2$
13. $4x^2 - 12xy + 9y^2$
15. $a^2 + 4a + 3$
17. $y^2 - 3y - 18$
19. $a^2 + 4a - 12$
21. $x^2 - 5x + 4$
23. $a^2 - 16$
25. $4x^2 - 9$
27. $x^2 + 4xy + 3y^2$
29. $x^2 - xy - 20y^2$
31. $4a^2 + 5ab - 6b^2$
33. $2a^2 - 10ab + 8b^2$
35. $6a^2 + ab - 12b^2$

37. $6x^2 + 23xy + 20y^2$

39. $4x^2 - 16xy + 15y^2$

***Problems 9-4***

1. $x^2 + 3x + 2$

3. $3x^2 + x + 2$

5. $4x^2 - 3x + 2$

7. $x + 3y$

9. $x - 3$

11. $x - 7$

13. $3x - 2$

15. $a - 5$

17. $2x + 3y$

19. $3a - 4b$

21. $x - 9y$

23. $8 + 4x$

25. $4x - 4 + \dfrac{3}{3x + 2}$

27. $2x + 3y - \dfrac{2xy}{4x + 5y}$

29. $4a - 2 + \dfrac{4}{a + 1}$

31. $6a + 3 + \dfrac{1}{a - 1}$

33. $a + b + \dfrac{2ab}{a + b}$

35. $6x^2 - 3x + 7 - \dfrac{3}{x + 1}$

***Problems 9-5***

1. $3(a + 3)$

3. $7(b - 3)$

5. $2xy(2x + 3)$

7. $2a(8a + 1)$

9. $8abc^2(3ac - 2b)$

11. $3a^2b^2(4a + 3b^2 - 2a^2b^2)$

13. $5xy^2(2x^2 - 4x^3y^2 + 3y)$

15. $3b^2c(5a^3c^3 - a^2bc^3 - 4)$

17. $5x^2y^2z^2(7x^2z + 5y^2z^2 - 6)$

19. $6x^2yz^2(4xyz^2 - 7y^2 - 12)$

21. $\dfrac{7x^2z^2}{3}\left(\dfrac{z}{4} - \dfrac{2xy^2}{3}\right)$

23. $\dfrac{6x^2y^2z}{5}\left(\dfrac{2x^2}{7} + \dfrac{4yz^2}{11} - \dfrac{3xyz^3}{5}\right)$

25. $\dfrac{7a^2c^2}{11b^2}\left(\dfrac{3a}{2} - \dfrac{4}{3b} - \dfrac{6ac}{4}\right)$

***Problems 9-6***

1. $(x + 2)(x + 3)$

3. $(x + 8)(x - 4)$

5. $(a + 3)(a + 3)$

7. $(a + 4)(a - 4)$

9. $(x + 7)(x - 7)$

11. $(x - 3)(x - 5)$

13. $(a - 5)(a - 7)$

15. $(a - 4)(a - 7)$

17. $(y - 4)(y + 9)$

19. $(x + 7)(x - 5)$

21. $(x + 4)(x - 6)$

23. $(2x + 3)(3x - 2)$

25. $(4a - 2)(2a - 4)$

27. $(3x + 4)(3x - 2)$

29. $(3b - 6)(2b - 2)$

31. $(5a + 3)(2a + 5)$

33. $(2x + 3y)(x + 2y)$

35. $(4a - 3b)(5a + 2b)$

37. $(2x + 6y)(4x - 2y)$

39. $(3x + 3y)(4x + 2y)$

# CHAPTER 10

***Problems 10-1***

1. $\frac{5x}{2} + 3 = 6$

   $5x + 6 = 12$

   $x = \frac{6}{5} = 1\frac{1}{5}$

3. $\frac{4a}{3} - 5 = 2$

   $4a - 15 = 6$

   $a = 5\frac{1}{4}$

5. $\frac{2I}{4} + \frac{I}{6} = 2$

   $6I + 2I = 24$

   $I = 3$

7. $\frac{2x}{3} - \frac{3x}{4} = 1$

   $8x - 9x = 23$

   $x = -12$

9. $\frac{4R}{2} = 3 + \frac{R}{3}$

   $12R = 18 + 2R$

   $R = 1\frac{4}{5}$

11. $\frac{R + 3}{4} - 2 = \frac{3R + 1}{2} + 4$

    $R + 3 - 2(4) = 2(3R+1) + 4(4)$

    $R + 3 - 8 = 6R + 2 + 16$

    $R = -4\frac{3}{5}$

13. $\frac{R - 2}{5} + 3 = \frac{2R + 2}{3} + 4$

    $3(R-2) + 3(15) = 5(2R+2) + 4(15)$
    $3R - 6 + 45 = 10R + 10 + 60$

    $R = -4\frac{3}{7}$

15. $\frac{c - 3}{4} - \frac{c + 1}{3} = 2 + 3c$

    $3(c-3) + 4(c+1) = 12(2+3c)$
    $3c - 9 - 4c - 4 = 24 + 36c$

    $c = -1$

17. $\frac{3}{I - 3} + 3 = 4$

    $3 + 3(I-3) = 4(I-3)$
    $3 + 3I - 9 = 4I - 12$

    $I = 6$

19. $\frac{4}{R - 2} + 2 = 3$

    $4 + 2(R-2) = 3(R-2)$
    $4 + 2R - 4 = 3R - 6$

    $R = 6$

21. $\frac{5x}{2} + \frac{x - 2}{6} = \frac{3x - 6}{4}$

    $6(5x) + 2(x-2) = 3(3x-6)$
    $30x + 2x - 4 = 9x - 18$
    $x = -\frac{14}{23}$

23. $\frac{4a}{3} + \frac{a + 2}{5} = \frac{2a - 3}{5}$

    $5(4a) + 3(a+2) = 3(2a-3)$
    $20a + 3a + 6 = 6a - 9$
    $a = -\frac{15}{17}$

25. $\frac{4 - x}{2x} - \frac{3}{x} = \frac{1}{x}$

    $4 - x - 3(2) = 1(2)$
    $4 - x - 6 = 2$
    $x = -4$

27. $\frac{3 - y}{2y} - \frac{2}{y} = \frac{3}{y}$

    $3 - y - 4 = 6$
    $y = -7$

29. $\frac{3}{G - 3} - \frac{1}{4} = 5$

    $3(4) - 1(G-3) = 5(4)(G-3)$
    $12 - 6 + 3 = 20G - 60$
    $G = 3\frac{4}{7}$

31. $\frac{2}{R + 3} - \frac{1}{3} = 4$

    $2(3) - 1(R+3) = 4(3)(R+3)$
    $6 - R - 3 = 12R + 36$
    $R = -2\frac{7}{13}$

33. $\frac{1}{a} + \frac{1}{b} = 2$

$b + a = 2ab$

*solve for a:*
$$2ab - a = b$$
$$a(2b-1) = b$$
$$a = \frac{b}{2b - 1}$$

*solve for b:*
$$2ab - b = a$$
$$b(2a-1) = a$$
$$b = \frac{a}{2a - 1}$$

35. $\frac{2}{a} - 3 = \frac{1}{b}$

$2b - 3a = a$

*solve for a:*
$$a + 3ab = 2b$$
$$a(1+3b) = 2b$$
$$a = \frac{2b}{1 + 3b}$$

*solve for b:*
$$b(2-3a) = a$$
$$b = \frac{3a}{2 - 3a}$$

37. $\frac{x}{2a} + \frac{x}{3b} = 2$

$3bx + 2ax = 12ab$

*solve for a:*
$$12ab - 2ax = 3bx$$
$$a(12b-2x) = 3bx$$
$$a = \frac{3bx}{12b - 2x}$$

*solve for b:*
$$12ab - 3bx = 2ax$$
$$b(12a-3x) = 2ax$$
$$b = \frac{2ax}{12a - 3x}$$

*solve for x:*
$$x(3b+2a) = 12ab$$
$$x = \frac{12ab}{2a + 3b}$$

39. $\frac{3x}{2a} - \frac{x}{3b} = 4$

$9bx - 2ax = 24ab$

*solve for a:*
$$24ab + 2ax = 9bx$$
$$a(24b+2x) = 9bx$$
$$a = \frac{9bx}{24b + 2x}$$

*solve for b:*
$$9bx - 24ab = 2ax$$
$$b(9x-24a) = 2ax$$
$$b = \frac{2ax}{9x - 24}$$

*solve for x:*
$$x(9b-2a) = 23ab$$
$$x = \frac{24ab}{9b - 2a}$$

41. $\frac{3}{a - 2} + \frac{2}{a + 2} = \frac{b}{a + 2}$

$3(a+2) + 2(a-2) = b(a-2)$

$3a + 6 + 2a - 4 = ab - 2b$

$5a + 2 = ab - 2b$

*solve for a:*
$$ab - 5a = 2b + 2$$
$$a(b-5) = 2b + 2$$
$$a = \frac{2b + 2}{b - 5}$$

*solve for b:*
$$ab - 2b = 5a + 2$$
$$b(a-2) = 5a + 2$$
$$b = \frac{5a + 2}{a - 2}$$

43. $\frac{2}{2a - 1} + \frac{3}{3a + 2} = \frac{2b}{2a - 1}$

$2(3a+2) + 3(2a-1) = 2b(3a+2)$

$6a + 4 + 6a - 3 = 6ab + 4b$

$12a + 1 = 6ab + 4b$

*solve for a:*
$$12a - 6ab = 4b - 1$$
$$a(12-6b) = 4b - 1$$
$$a = \frac{4b - 1}{12 - 6b}$$

*solve for b:*
$$b(6a+4) = 12a + 1$$
$$b = \frac{12a + 1}{6a + 4}$$

45. $\frac{5}{3b - 2} - \frac{4}{3b - 1} = \frac{3}{3b - 1}$

$5(3b-1) - 4(3b-2) = 3(3b-2)$

$15b - 5 - 12b + 8 = 9b - 6$

$$b = 1\frac{1}{2}$$

***Problems 10-2***

1. $V_L = \dfrac{V_{OC}R_L}{R_{TH} + R_L}$

$V_LR_{TH} + V_LR_L = V_{OC}R_L$

$V_{OC} = \dfrac{V_LR_{TH} + V_LR_L}{R_L}$

$R_{TH} = \dfrac{V_{OC}R_L - V_LR_L}{V_L}$

$R_L = \dfrac{V_LR_{TH}}{V_{OC} - V_L}$

3. $I_L = \dfrac{I_{SC}G_L}{G_L + G_N}$

$I_LG_L + I_LG_N = I_{SC}G_L$

$I_{SC} = \dfrac{I_LG_L + I_LG_N}{G_L}$

$G_L = \dfrac{I_LG_N}{I_{SC} - I_L}$

$G_N = \dfrac{I_{SC}G_L - I_LG_L}{I_L}$

5. $R_s = \dfrac{R_iR_o}{R_i + R_o}$

$R_sR_i + R_sR_o = R_iR_o$

$R_o = \dfrac{R_sR_i}{R_i - R_s}$

$R_i = \dfrac{R_sR_o}{R_o - R_s}$

7. $r_i = \beta(r_e + \dfrac{r_b}{\beta})$

$r_i = \beta r_e + r_b$

$r_e = \dfrac{r_i - r_b}{\beta}$

$r_b = r_i - \beta r_e$

9. $R_T = \dfrac{R_1R_2}{R_1 + R_2}$

$R_TR_1 + R_TR_2 = R_1R_2$

$R_1 = \dfrac{R_TR_2}{R_2 - R_T}$

$R_2 = \dfrac{R_TR_1}{R_1 - R_T}$

11. $\alpha(1 + \dfrac{1}{\alpha}) = B$

$\alpha = 1 = B$

$\alpha = B - 1$

13. $2\pi fC(R_1 + R_2) = 1$

$2\pi fCR_1 + 2\pi fCR_2 = 1$

$R_1 = \dfrac{1 - 2\pi fCR_2}{2\pi fC}$

15. $I_B(R_1+R_2) = V_{CC} - V_B$

$V_B = V_{CC} - I_B(R_1+R_2)$

$I_BR_1 + R_1R_2 = V_{CC} - V_B$

$R_1 = \dfrac{V_{CC} - I_BR_2}{I_B}$

***Problems 10-3***

1. $V_2(R_1+R_2) = ER_2$

   $V_2R_1 + V_2R_2 = ER_2$

   a) $R_2 = \dfrac{V_2R_1}{E - V_2} = 4.8\ k\Omega$

   b) $R_1 = \dfrac{ER_2 - V_2R_2}{V_2}$

   $R_1 = 773\ \Omega$

3. $I_1 = \dfrac{I_TG_1}{G_1 + G_2}$

   $I_1G_1 + I_1G_2 = I_TG_1$

   a) $I_T = \dfrac{I_1G_1 - I_1G_1}{G_1} = 12\ mA$

   b) $G_2 = \dfrac{I_TG_1 - I_1G_1}{I_1} = 1.79\ mS$

5. $A_i + A_ih_{oe}R_L = h_{fe}$

   a) $h_{oe} = \dfrac{h_{fe} - A_i}{A_iR_L} = 74.3\ \mu S$

   b) $R_L = \dfrac{h_{fe} - A_i}{A_ih_{oe}} = 20.8\ k\Omega$

7. $R_T = \dfrac{R_1R_2}{R_1 + R_2}$

   $R_1 = \dfrac{R_2R_T}{R_2 - R_T} = 3\ k\Omega$

9. $Z_T = \dfrac{Z_1Z_2}{Z_1 + Z_L}$

   $Z_2 = \dfrac{Z_1Z_T}{Z_1 - Z_2} = 220\ k\Omega$

11. $I_BR_1 + I_BR_2 = V_{CC} - V_B$

    $R_2 = \dfrac{V_{CC} - V_B - I_BR_1}{I_B} = 3.94\ k\Omega$

13. $I_C = \dfrac{\beta V_{CC}}{\beta R_E + R_B}$

    $\beta I_CR_E + I_CR_B = \beta V_{CC}$

    $R_B = \dfrac{\beta V_{CC} - \beta I_CR_E}{I_C} = 550\ k\Omega$

15. $2\pi fCR_1 + 2\pi fCR_2$

    $R_1 = \dfrac{1 - 2\pi fCR_2}{2\pi fC} = 9.34\ k\Omega$

***Problems 10-4***

1. $(x - 4)(x + 4) = 0$
   $x - 4 = 0,\ x + 4 = 0$
   $x = 4,\ x = -4$

3. $x^2 + 6x + 8 = 0$
   $(x + 2)(x + 4) = 0$
   $x = -2,\ x = -4$

5. $x^2 + 3x - 18 = 0$
   $(x + 6)(x - 3) = 0$
   $x + 6 = 0,\ x - 3 = 0$
   $x = -6,\ x = 3$

7. $x^2 - 6x + 8 = 0$
   $(x - 4)(x - 2) = 0$
   $x - 4 = 0,\ x - 2 = 0$
   $x = 4,\ x = 2$

9. $x^2 - x - 56 = 0$
   $(x + 7)(x - 8) = 0$
   $x + 7 = 0,\ x - 8 = 0$
   $x = -7,\ x = 8$

11. $x^2 - 3x - 40 = 0$
    $(x + 5)(x - 8) = 0$
    $x + 5 = 0,\ x - 8 = 0$
    $x = -5,\ x = 8$

13. $x^2 - 13x + 36 = 0$
$(x - 9)(x - 4) = 0$
$x - 9 = 0,\ x - 4 = 0$
$x = 9,\ x = 4$

15. $x^2 - 3x - 18 = 0$
$(x + 3)(x - 6) = 0$
$x + 3 = 0,\ x - 6 = 0$
$x = -3,\ x = 6$

17. $x^2 + 10x + 21 = 0$
$(x + 3)(x + 7) = 0$
$x + 3 = 0,\ x + 7 = 0$
$x = -3,\ x = -7$

19. $x^2 - 3x + 2 = 0$
$(x - 2)(x - 1) = 0$
$x - 2 = 0,\ x - 1 = 0$
$x = 2,\ x = 1$

***Problems 10-5***

1. $x^2 + 8x + 12 = 0$
$(x + 2)(x + 6) = 0$
$x + 2 = 0,\ x + 6 = 0$
$x = -2,\ x = -6$

3. $a = 1,\ b = 6,\ c = 6$
$$x = \frac{-6 \pm [6^2 - 4(1)(6)]^{1/2}}{2(1)} = \frac{-6 \pm 3.46}{2}$$
$x = -4.73,\ x = -1.27$

5. $a = 1,\ b = 2,\ c = -14$
$$x = \frac{-2 \pm [2^2 - 4(1)(-14)]^{1/2}}{2(1)} = \frac{-2 \pm 7.75}{2}$$
$x = -4.87,\ x = 2.87$

7. $a = 1,\ b = 15,\ c = -15$
$$x = \frac{-(-15) \pm [-15^2 - 4(1)(-15)]^{1/2}}{2(1)} = \frac{15 \pm 16.9}{2}$$
$x = 15.9,\ x = -0.941$

9. $x^2 - 7x - 4 = 0$
$a = 1,\ b = -7,\ c = -4$
$$x = \frac{-(-7) \pm [-7^2 - 4(1)(-4)]^{1/2}}{2(1)} = \frac{7 \pm 8.06}{2}$$
$x = 7.53,\ x = -0.531$

11. $a = 3,\ b = 2,\ c = -1$
$$x = \frac{-2 \pm [2^2 - 4(3)(-1)]^{1/2}}{2(3)} = \frac{-2 \pm 4}{6}$$
$x = -1,\ x = 0.333$

13. $a = 2,\ b = -5,\ c = -5$
$$x = \frac{-(-5) \pm [-5^2 - 4(2)(-5)]^{1/2}}{2(2)} = \frac{5 \pm 8.06}{4}$$
$x = 3.27,\ x = -0.766$

15. $a = 4, b = 12, c = 8$

$$x = \frac{-12 +/- [12^2 - 4(4)(8)]^{1/2}}{2(4)} = \frac{-12 +/- 4}{8}$$

$x = -2, x = -1$

17. $3x^2 + 7x + 3 = 0$

$a = 3, b = 7, c = 3$

$$x = \frac{-7 +/- [7^2 - 4(3)(3)]^{1/2}}{2(3)} = \frac{-7 +/- 3.61}{6}$$

$x = -1.77, x = -0.566$

19. $3x^2 - 14x - 20 = 0$

$a = 3, b = -14, c = -20$

$$x = \frac{-(-14) +/- [-14^2 - 4(3)(-20)]^{1/2}}{2(3)} = \frac{14 +/- 20.9}{6}$$

$x = 5.81, x = -1.15$

# CHAPTER 11

***Problems 11-1***

***In problems 1-8:*** $I_T = I_1 + I_2 + I_3$

1. $I_T = I_1 + I_2 + I_3 = 3\text{ A} + 2\text{ A} + 4\text{ A} = 9\text{ A}$
3. $I_3 = I_T - (I_1 + I_2) = 20\text{ mA} - (6\text{ mA} + 11\text{ mA}) = 3\text{ mA}$
5. $I_2 = I_T - (I_1 + I_3) = 65\ \mu\text{A} - (10\ \mu\text{A} + 10\ \mu\text{A}) = 45\ \mu\text{A}$
7. $I_1 = I_T - (I_2 + I_3) = 1.3\text{ A} - (300\text{ mA} + 40\text{ mA}) = 960\text{ mA}$

***In problems 9-14:*** $I_2 = I_3 \quad I_T = I_1 + I_2 = I_1 + I_3$

9. $I_T = I_1 + I_2 = 10\text{ mA} + 15\text{ mA} = 25\text{ mA}$
   $I_3 = I_2 = 15\text{ mA}$
11. $I_1 = I_T - I_3 = 1.6\text{ A} - 800\text{ mA} = 800\text{ mA}$
   $I_2 = I_3 = 800\text{ mA}$
13. $I_2 = I_T - I_1 = 600\ \mu\text{A} - 400\ \mu\text{A} = 200\ \mu\text{A}$
   $I_3 = I_2 = 200\ \mu\text{A}$

***In problems 15-18:*** $I_T = I_1 = I_4 + I_2 = I_4 + I_3$

15. $I_T = I_1 = 1.6\text{ A};\ I_4 = I_1 - I_2 = 1.6\text{ A} - 800\text{ mA} = 800\text{ mA}$
   $I_3 = I_2 = 800\text{ mA}$
17. $I_2 = I_3 = 500\ \mu\text{A};\ I_T = I_1 = I_3 + I_4 = 500\ \mu\text{A} + 2.7\text{ mA} = 3.2\text{ mA}$

***In problems 19-22:*** $I_T = I_1 + I_2 + I_3 = I_1 + I_4;\ I_2 + I_3 = I_4$

19. $I_1 = I_T - I_4 = 6.7\text{ mA} - 5\text{ mA} = 1.7\text{ mA}$
   $I_3 = I_4 - I_2 = 5\text{ mA} - 1.2\text{ mA} = 3.8\text{ mA}$
21. $I_T = I_1 + I_4 = 6\text{ mA} + 8.2\text{ mA} = 14.2\text{ mA}$
   $I_2 = I_4 - I_3 = 8.2\text{ mA} - 3.4\text{ mA} = 4.8\text{ mA}$
23. $I_T = I_6 + I_1 = I_6 + I_5 = I_6 + I_2 + I_3 = I_6 + I_2 + I_4$
   $I_1 = I_5 = 30\text{ mA};\ I_6 = I_T - I_5 = 100\text{ mA} - 30\text{ mA} = 70\text{ mA}$
   $I_3 = I_4 = I_1 - I_2 = 30\text{ mA} - 20\text{ mA} = 10\text{ mA}$

***Problems 11-2***

***In problems 1-4:*** $E = V_1 + V_2$

1. $V_2 = E - V_1 = 30\text{ V} - 12\text{ V} = 18\text{ V}$
3. $E = V_1 + V_2 = 7\text{ V} + 6\text{ V} = 13\text{ V}$

***In problems 5-10:*** $E = V_1 + V_2 + V_3$

5. $V_3 = E - (V_1 + V_2) = 70\text{ V} - (22\text{ V} + 37\text{ V}) = 11\text{ V}$
7. $E = V_1 + V_2 + V_3 = 3\text{ V} + 8\text{ V} + 7\text{ V} = 18\text{ V}$
9. $V_1 = E - (V_2 + V_3) = 37\text{ V} - (20\text{ V} + 3\text{ V}) = 14\text{ V}$

***In problems 11-16:*** $E = V_1 + V_2 = V_1 + V_3;\ V_2 = V_3$

11. $V_2 = V_3 = 8\text{ V};\ V_1 = E - V_3 = 20\text{ V} - 8\text{ V} = 12\text{ V}$
13. $V_2 = V_3 = E - V_1 = 18\text{ V} - 5\text{ V} = 13\text{ V}$
15. $E = V_1 + V_3 = 12\text{ V} + 5\text{ V} = 17\text{ V};\ V_2 = V_3 = 5\text{ V}$

*In problems 17-22:* $E = V_1 + V_4 = V_1 + V_2 + V_3 = V_4 = V_2 + V_3$

17. $V_4 = V_2 + V_3 = 3\text{ V} + 6\text{ V} = 9\text{ V}$; $V_1 = E - V_4 = 12\text{ V} - 9\text{ V} = 3\text{ V}$

19. $V_3 = V_4 - V_2 = 1.3\text{ V} - 500\text{ mV} = 800\text{ mV}$; $V_1 = E - V_4 = 3\text{ V} - 1.3\text{ V} = 1.7\text{ V}$

21. $E = V_1 + V_4 = 20\text{ V} + 40\text{ V} = 60\text{ V}$; $V_3 = V_4 - V_2 = 40\text{ V} - 30\text{ V} = 10\text{ V}$

*In problems 23-26:* $E = V_6$; $E = V_1 + V_2 + V_5 = V_1 + V_2 + V_3 + V_4$; $V_2 = V_3 + V_4$

23. $V_2 = E - (V_1 + V_5) = 50\text{ V} - (10\text{ V} + 15\text{ V}) = 25\text{ V}$

    $V_4 = V_2 - V_3 = 25\text{ V} - 8\text{ V} = 17\text{ V}$

    $V_6 = E = 50\text{ V}$

25. $E = V_6 = 25\text{ V}$; $V_2 = E - (V_1 + V_5) = 25\text{ V} - (9\text{ V} + 4\text{ V}) = 12\text{ V}$

    $V_3 = V_2 - V_4 = 12\text{ V} - 6.5\text{ V} = 5.5\text{ V}$

***Problems 11-3***

1. a) $V_2 = V_3 - V_1$
   $= 25\text{ V} - 15\text{ V} = 10\text{ V}$
   b) $V_4 = E - V_3 = 30\text{ V} - 25\text{ V}$
   $= 5\text{ V}$
   c) 25 V positive
   d) 30 V positive
   e) 5 V negative

3. a) $V_1 = E_1 - V_2 = 20\text{ V} - 8\text{ V}$
   $= 12\text{ V}$
   b) $V_3 = E_2 + V_2 = 35\text{ V} + 8\text{ V}$
   $= 43\text{ V}$
   c) 43 V negative
   d) 35 V negative
   e) 8 V negative

5. a) $V_3 = E_1 - (V_1 + V_2)$
   $= 25\text{ V} - (6\text{ V} + 8\text{ V}) = 11\text{ V}$
   b) $V_4 = V_3 - E_2 = 11\text{ V} - 6\text{ V}$
   $= 5\text{ V}$
   c) 19 V positive
   d) 13 V negative

7. a) $V_1 = E_1 - V_3 = 40\text{ V} - 6\text{ V}$
   $= 34\text{ V}$
   b) $V_2 = E_2 + V_3 = 10\text{ V} + 6\text{ V}$
   $= 16\text{ V}$
   c) 6 V positive
   d) 50 V negative

# CHAPTER 12

***Problems 12-1***

***In the following problems:*** $R_T = R_1 + R_2$

1. a) $R_T$ = 45 kΩ
   b) $R_T$ = 1.95 kΩ
   c) $R_T$ = 16.8 kΩ
   d) $R_T$ = 300 kΩ

3. a) $R_2$ = 1.8 kΩ
   b) $R_2$ = 39 kΩ

***In the following problems:*** $R_T = R_1 + R_2 + R_3$

5. a) $R_T$ = 2.29 kΩ
   b) $R_T$ = 130 kΩ
   c) $R_T$ = 297 Ω
   d) $R_T$ = 1.36 MΩ

7. a) $R_3$ = 5.6 kΩ
   b) $R_3$ = 82 kΩ

***In the following problems:*** $G_T = G_1 + G_2$ $\qquad R_T = \frac{1}{G_T}$

9. a) $G_T$ = 150 μS
   $R_T$ = 6.67 kΩ
   b) $G_T$ = 1.20 mS
   $R_T$ = 831 Ω
   c) $G_T$ = 76.8 μS
   $R_T$ = 13.0 kΩ
   d) $G_T$ = 20 μS
   $R_T$ = 50.0 kΩ

***In the following problems:*** $G_T = G_1 + G_2 + G_3$ $\qquad R_T = \frac{1}{G_T}$

11. a) $G_T$ = 685 μS
    $R_T$ = 1.46 kΩ
    b) $G_T$ = 31.9 μS
    $R_T$ = 31.3 kΩ
    c) $G_T$ = 2.39 mS
    $R_T$ = 419 Ω
    d) $G_T$ = 26.7 μS
    $R_T$ = 37.4 kΩ

13. a) $G_2 = G_T - (G_1+G_3) = 261\ \mu S - (48.2\ \mu S) = 213\ \mu S$

$$R_2 = \frac{1}{G_2} = \frac{1}{213\ \mu S} = 4.7\ k\Omega$$

b) $G_2 = G_T - (G_1+G_2) = 18.4\ mS - (13.3\ mS) = 5.02\ mS$

$$R_2 = \frac{1}{G_2} = \frac{1}{5.02\ mS} = 199\ \Omega$$

***Problems 12-2***

*In the following problems:* $R_T = R_1 + R_2 /\!/ R_3$

1. a) $R_T = R_1 = \frac{1}{G_1 + G_2} = 560\ \Omega + \frac{1}{2.22\ mS} = 1.01\ k\Omega$

b) $R_T = R_1 = \frac{1}{G_1 + G_2} = 7.5\ k\Omega + \frac{1}{92.6\ \mu S} = 18.3\ k\Omega$

c) $R_T = R_1 = \frac{1}{G_1 + G_2} = 27\ k\Omega + \frac{1}{105\ \mu S} = 36.6\ k\Omega$

d) $R_T = R_1 = \frac{1}{G_1 + G_2} = 330\ k\Omega + \frac{1}{2.47\ \mu S} = 735\ k\Omega$

*In the following problems:* $G_X = \frac{1}{R_1 + R_2}$; $R_T = (R_1+R_2) /\!/ R_3$

3. a) $R_T = \frac{1}{G_X + G_3} = \frac{1}{87\ \mu S + 110\ \mu S} = 5.08\ k\Omega$

b) $R_T = \frac{1}{G_X + G_3} = \frac{1}{303\ \mu S + 1.22\ mS} = 657\ \Omega$

c) $R_T = \frac{1}{G_X + G_3} = \frac{1}{22.2\ \mu S + 17.9\ \mu S} = 25.0\ k\Omega$

d) $R_T = \frac{1}{G_X + G_3} = \frac{1}{870\ \mu S + 455\ \mu S} = 755\ \Omega$

*In the following problems:* $G_X = G_2 + G_3 + G_4$ *and* $R_T = R_1 + \frac{1}{G_X}$

5. a) $R_T = R_1 + \frac{1}{G_X} = 12\ k\Omega + \frac{1}{88.6\ \mu S} = 23.3\ k\Omega$

b) $R_T = R_1 + \frac{1}{G_X} = 120\ \Omega + \frac{1}{4.91\ mS} = 324\ \Omega$

c) $R_T = R_1 + \frac{1}{G_X} = 390\ k\Omega + \frac{1}{3.39\ \mu S} = 685\ k\Omega$

5. d) $R_T = R_1 + \frac{1}{G_X} = 5.1\ k\Omega + \frac{1}{347\ \mu S} = 7.98\ k\Omega$

7. $R_X = R_2 /\!/_3 = 6\ k\Omega$

$R_Y = R_5 /\!/ R_6 = 2.56\ k\Omega$

$R_Z = R_X + R_Y + R_4 = 6\ k\Omega + 2.56\ k\Omega = 6.8\ k\Omega = 15.4\ k\Omega$

$R_T = R_1 /\!/ R_Z = 1.11\ k\Omega$

9. $R_X = R_1 /\!/ {}_2 + R_3 = 83.1\ k\Omega + 91\ k\Omega = 174\ k\Omega$

$R_Y = R_5 /\!/ R_6 /\!/ R_7 + R_8 /\!/ R_9 = 204\ k\Omega + 372\ k\Omega = 575\ k\Omega$

$R_T = R_X /\!/ R_Y /\!/ R_4 = 174\ k\Omega /\!/ 575\ k\Omega /\!/ 330\ k\Omega = 95.1\ k\Omega$

***Problems 12-3***

1. a) $R_T = R_1 + R_2 = 950\ \Omega,\ I = \frac{E}{R_T} = \frac{9\ V}{950\ \Omega} = 9.47\ mA$

$V_1 = IR_1 = 2.56\ V,\ V_2 = IR_2 = 6.44\ V$

b) $R_T = R_1 + R_2 = 4.5\ k\Omega,\ I = \frac{E}{R_T} = \frac{12\ V}{4.5\ k\Omega} = 2.67\ mA$

$V_1 = IR_1 = 3.2\ V,\ V_2 = IR_2 = 8.80\ V$

c) $R_T = R_1 + R_2 = 340\ k\Omega,\ I = \frac{E}{R_T} = \frac{40\ V}{340\ k\Omega} = 118\ \mu A$

$V_1 = IR_1 = 14.1\ V,\ V_2 = IR_2 = 25.9\ V$

d) $R_T = R_1 + R_2 = 13.3\ k\Omega,\ I = \frac{E}{R_T} = \frac{10\ V}{13.3\ k\Omega} = 752\ \mu A$

$V_1 = IR_1 = 7.52\ V,\ V_2 = IR_2 = 2.48\ V$

3. a) $V_1 = E - V_2 = 10\ V - 3\ V = 7\ V,\ I = \frac{V_1}{R_1} = \frac{7\ V}{1\ k\Omega} = 7\ mA$

$R_T = \frac{E}{I} = \frac{10\ V}{7\ mA} = 1.43\ k\Omega,\ R_2 = R_T - R_1 = 429\ \Omega$

b) $V_1 = E - V_2 = 40\ V - 23.5\ V = 16.5\ V,\ I = \frac{V_1}{R_1} = \frac{16.5\ V}{6.8\ k\Omega} = 2.43\ mA$

$R_T = \frac{E}{I} = \frac{40\ V}{2.43\ mA} = 16.5\ k\Omega,\ R_2 = R_T - R_1 = 9.68\ k\Omega$

c) $V_1 = E - V_2 = 15\text{ V} - 12\text{ V} = 3\text{ V}$, $I = \dfrac{V_1}{R_1} = \dfrac{3\text{ V}}{8.2\text{ k}\Omega} = 366\ \mu\text{A}$

$R_T = \dfrac{E}{I} = \dfrac{15\text{ V}}{366\ \mu\text{A}} = 41\text{ k}\Omega$, $R_2 = R_T - R_1 = 32.8\text{ k}\Omega$

d) $V_1 = E - V_2 = 100\text{ V} - 29.3\text{ V} = 70.7\text{ V}$, $I = \dfrac{V_1}{R_1} = \dfrac{70.7\text{ V}}{220\text{ k}\Omega} = 321\ \mu\text{A}$

$R_T = \dfrac{E}{I} = \dfrac{100\text{ V}}{320\ \mu\text{A}} = 311\text{ k}\Omega$, $R_2 = R_T - R_1 = 91.2\text{ k}\Omega$

5. a) $V_2 = IR_2 = 213\ \mu\text{A} \times 33\text{ k}\Omega = 7.03\text{ V}$, $V_1 = E - V_2 = 13\text{ V}$

$R_1 = \dfrac{V_1}{I} = \dfrac{13\text{ V}}{213\ \mu\text{A}} = 60.9\text{ k}\Omega$, $R_T = R_1 + R_2 = 93.9\text{ k}\Omega$

b) $V_2 = IR_2 = 1\text{ mA} \times 10\text{ k}\Omega = 10\text{ V}$, $V_1 = E - V_2 = 15\text{ V}$

$R_1 = \dfrac{V_1}{I} = \dfrac{15\text{ V}}{1\text{ mA}} = 15\text{ k}\Omega$, $R_T = R_1 + R_2 = 25\text{ k}\Omega$

c) $V_2 = IR_2 = 7.69\text{ mA} \times 1\text{ k}\Omega = 7.69\text{ V}$, $V_1 = E - V_2 = 4.31\text{ V}$

$R_1 = \dfrac{V_1}{I} = \dfrac{4.31\text{ V}}{7.69\text{ mA}} = 560\ \Omega$, $R_T = R_1 + R_2 = 1.56\text{ k}\Omega$

d) $V_2 = IR_2 = 20\ \mu\text{A} \times 180\text{ k}\Omega = 3.6\text{ V}$, $V_1 = E - V_2 = 5.4\text{ V}$

$R_1 = \dfrac{V_1}{I} = \dfrac{5.4\text{ V}}{20\ \mu\text{A}} = 270\text{ k}\Omega$, $R_T = R_1 + R_2 = 450\text{ k}\Omega$

7. a) $V_2 = IR_2 = 75\ \mu\text{A} \times 3\text{ k}\Omega = 225\text{ mV}$, $E = V_1 + V_2 = 231\text{ mV}$

$R_T = \dfrac{E}{I} = \dfrac{231\text{ mV}}{75\ \mu\text{A}} = 3.08\text{ k}\Omega$, $R_1 = R_T - R_2 = 82.7\ \Omega$

b) $V_2 = IR_2 = 2\text{ mA} \times 12\text{ k}\Omega = 24\text{ V}$, $E = V_1 + V_2 = 38.3\text{ V}$

$R_T = \dfrac{E}{I} = \dfrac{38.3\text{ V}}{2\text{ mA}} = 19.2\text{ k}\Omega$, $R_1 = R_T - R_2 = 7.15\text{ k}\Omega$

c) $V_2 = IR_2 = 167\ \mu\text{A} \times 33\text{ k}\Omega = 5.50\text{ V}$, $E = V_1 + V_2 = 10\text{ V}$

$R_T = \dfrac{E}{I} = \dfrac{4.56\text{ V}}{167\ \mu\text{A}} = 60\text{ k}\Omega$, $R_1 = R_T - R_2 = 27\text{ k}\Omega$

d) $V_2 = IR_2 = 14.9\text{ mA} \times 82\ \Omega = 1.22\text{ V}$, $E = V_1 + V_2 = 3\text{ V}$

$R_T = \dfrac{E}{I} = \dfrac{3\text{ V}}{14.9\text{ mA}} = 202\ \Omega$, $R_1 = R_T - R_2 = 120\ \Omega$

9. a) $R_T = R_1 + R_2 + R_3 = 61\ k\Omega$, $I = \dfrac{E}{R_T} = \dfrac{9\ V}{61\ k\Omega} = 148\ \mu A$

$V_1 = IR_1 = 4.87\ V$, $V_2 = IR_2 = 1.48\ V$, $V_3 = IR_3 = 2.66\ V$

b) $R_T = R_1 + R_2 + R_3 = 388\ k\Omega$, $I = \dfrac{E}{R_T} = \dfrac{40\ V}{388\ k\Omega} = 103\ \mu A$

$V_1 = IR_1 = 22.7\ V$, $V_2 = IR_2 = 10.3\ V$, $V_3 = IR_3 = 7.01\ V$

c) $R_T = R_1 + R_2 + R_3 = 2.58\ k\Omega$, $I = \dfrac{E}{R_T} = \dfrac{25\ V}{2.58\ k\Omega} = 9.69\ mA$

$V_1 = IR_1 = 7.95\ V$, $V_2 = IR_2 = 11.6\ V$, $V_3 = IR_3 = 5.43\ V$

d) $R_T = R_1 + R_2 + R_3 = 8.6\ k\Omega$, $I = \dfrac{E}{R_T} = \dfrac{15\ V}{8.6\ k\Omega} = 1.74\ mA$

$V_1 = IR_1 = 3.14\ V$, $V_2 = IR_2 = 7.5\ V$, $V_3 = IR_3 = 4.36\ V$

11. a) $V_2 = IR_2 = 1\ mA \times 10\ k\Omega = 10\ V$, $V_3 = E - (V_1+V_2) = 10\ V$

$R_1 = \dfrac{V_1}{I} = \dfrac{5\ V}{1\ mA} = 5\ k\Omega$, $R_3 = \dfrac{V_3}{I} = 10\ k\Omega$, $R_T = R_1 + R_2 + R_3 = 25\ k\Omega$

b) $V_2 = IR_2 = 150\ \mu A \times 47\ k\Omega = 7.05\ V$, $V_3 = E - (V_1+V_2) = 6.25\ V$

$R_1 = \dfrac{V_1}{I} = 44.7\ k\Omega$, $R_3 = \dfrac{V_3}{I} = 41.7\ k\Omega$, $R_T = R_1 + R_2 + R_3 = 133\ k\Omega$

c) $V_2 = IR_2 = 1.66\ mA \times 8.2\ k\Omega = 13.6\ V$, $V_3 = E - (V_1+V_2) = 36.5\ V$

$R_1 = \dfrac{V_1}{I} = 18\ k\Omega$, $R_3 = \dfrac{V_3}{I} = 22\ k\Omega$, $R_T = R_1 + R_2 + R_3 = 48.2\ k\Omega$

d) $V_2 = IR_2 = 15.5\ mA \times 91\ \Omega = 1.41\ V$, $V_3 = E - (V_1+V_2) = 1.16\ V$

$R_1 = \dfrac{V_1}{I} = 27.9\ \Omega$, $R_3 = \dfrac{V_3}{I} = 74.6\ \Omega$, $R_T = R_1 + R_2 + R_3 = 194\ \Omega$

13. a) $E = IR_T = 3\ mA \times 12\ k\Omega = 36\ V$, $V_1 = IR_1 = 14.1\ V$

$V_3 = E - (V_1+V_2) = 14.6\ V$, $R_2 = \dfrac{V_2}{I} = 2.43\ k\Omega$, $R_3 = \dfrac{V_3}{I} = 4.87\ k\Omega$

b) $E = IR_T = 270\ \mu A \times 43.2\ k\Omega = 11.7\ V$, $V_1 = IR_1 = 4.05\ V$

$V_3 = E - (V_1+V_2) = 3.95\ V$, $R_2 = \dfrac{V_2}{I} = 13.7\ k\Omega$, $R_3 = \dfrac{V_3}{I} = 14.6\ k\Omega$

c) $E = IR_T = 66.7\ \mu A \times 450\ k\Omega = 30\ V$, $V_1 = IR_1 = 13.3\ V$

$V_3 = E - (V_1+V_2) = 10\ V$, $R_2 = \dfrac{V_2}{I} = 100\ k\Omega$, $R_3 = \dfrac{V_3}{I} = 150\ k\Omega$

d) $E = IR_T = 2.3\text{ mA} \times 870\ \Omega = 2\text{ V},\ V_1 = IR_1 = 1.08\text{ V}$

$V_3 = E - (V_1+V_2) = 229\text{ mV},\ R_2 = \dfrac{V_2}{I} = 300\ \Omega,\ R_3 = \dfrac{V_3}{I} = 100\ \Omega$

***Problems 12-4***

1. a) $G_T = G_1 + G_2 = 3.7\text{ mS} + 1.79\text{ mS} = 5.49\text{ mS},\ R_T = \dfrac{1}{G_T} = 182\ \Omega$

$V = \dfrac{I_T}{G_T} = \dfrac{3\text{ mA}}{5.49\text{ mS}} = 547\text{ mV},\ I_1 = VG_1 = 2.02\text{ mA},\ I_2 = VG_2 = 977\ \mu A$

b) $G_T = G_1 + G_2 = 14.7\ \mu S + 25.6\ \mu S = 40.3\ \mu S,\ R_T = \dfrac{1}{G_T} = 24.8\text{ k}\Omega$

$V = \dfrac{I_T}{G_T} = \dfrac{700\ \mu A}{40.3\ \mu S} = 17.3\text{ V},\ I_1 = VG_1 = 255\ \mu A,\ I_2 = VG_2 = 444\ \mu A$

c) $G_T = G_1 + G_2 = 4.55\text{ mS} + 3.7\text{ mS} = 8.25\text{ mS},\ R_T = \dfrac{1}{G_T} = 121\ \Omega$

$V = \dfrac{I_T}{G_T} = \dfrac{37.6\text{ mA}}{8.25\text{ mS}} = 4.56\text{ V},\ I_1 = VG_1 = 20.7\text{ mA},\ I_2 = VG_2 = 16.9\text{ mA}$

d) $G_T = G_1 + G_2 = 17.9\ \mu S + 30.3\ \mu S = 48.2\ \mu S,\ R_T = \dfrac{1}{G_T} = 20.8\text{ k}\Omega$

$V = \dfrac{I_T}{G_T} = \dfrac{100\ \mu A}{48.2\ \mu S} = 2.08\text{ V},\ I_1 = VG_1 = 37.1\ \mu A,\ I_2 = VG_2 = 62.9\ \mu A$

3. a) $I_1 = I_1 - I_2 = 10\text{ mA},\ V = I_1R_1 = 6.8\text{ V},\ R_2 = \dfrac{V}{I_2} = 453\ \Omega$

$G_T = \dfrac{I_T}{V} = \dfrac{25\text{ mA}}{6.8\text{ V}} = 3.68\text{ mS},\ R_T = \dfrac{1}{G_T} = 272\ \Omega$

b) $I_1 = I_1 - I_2 = 4\text{ mA},\ V = I_1R_1 = 7.2\text{ V},\ R_2 = \dfrac{V}{I_2} = 7.2\text{ k}\Omega$

$G_T = \dfrac{I_T}{V} = \dfrac{5\text{ mA}}{7.2\text{ V}} = 694\ \mu S,\ R_T = \dfrac{1}{G_T} = 1.44\text{ k}\Omega$

c) $I_1 = I_1 - I_2 = 1.25\text{ mA},\ V = I_1R_1 = 5.88\text{ V},\ R_2 = \dfrac{V}{I_2} = \dfrac{5.88\text{ V}}{1.05\text{ mA}} = 5.6\text{ k}\Omega$

$G_T = \dfrac{I_T}{V} = \dfrac{2.3\text{ mA}}{5.88\text{ V}} = 391\ \mu S,\ R_T = \dfrac{1}{G_T} = 2.56\text{ k}\Omega$

d) $I_1 = I_1 - I_2 = 17.7\text{ mA}$, $V = I_1R_1 = 83.4\text{ V}$, $R_2 = \dfrac{V}{I_2} = 6.8\text{ k}\Omega$

$G_T = \dfrac{I_T}{V} = \dfrac{30\text{ mA}}{83.2\text{ V}} = 360\ \mu\text{S}$, $R_T = \dfrac{1}{G_T} = 2.78\text{ k}\Omega$

5. a) $G_T = \dfrac{I_T}{V} = \dfrac{10\text{ mA}}{20\text{ V}} = 500\ \mu\text{S}$, $R_T = \dfrac{1}{G_T} = 2\text{ k}\Omega$

$I_1 = \dfrac{V}{R_1} = 4.26\text{ mA}$, $I_2 = I_T - I_1 = 5.74\text{ mA}$, $R_2 = \dfrac{V}{I_2} = 3.48\text{ k}\Omega$

b) $G_T = \dfrac{I_T}{V} = \dfrac{350\ \mu\text{A}}{25\text{ V}} = 14.0\ \mu\text{S}$, $R_T = \dfrac{1}{G_T} = 71.4\text{ k}\Omega$

$I_1 = \dfrac{V}{R_1} = 139\ \mu\text{A}$, $I_2 = I_T - I_1 = 211\ \mu\text{A}$, $R_2 = \dfrac{V}{I_2} = 118\text{ k}\Omega$

c) $G_T = \dfrac{I_T}{V} = \dfrac{20\text{ mA}}{9.74\text{ V}} = 2.05\text{ mS}$, $R_T = \dfrac{1}{G_T} = 488\ \Omega$

$I_1 = \dfrac{V}{R_1} = \dfrac{9.74\text{ V}}{820\ \Omega} = 11.9\text{ mA}$, $I_2 = I_T - I_1 = 8.12\text{ mA}$, $R_2 = \dfrac{V}{I_2} = 1.2\text{ k}\Omega$

d) $G_T = \dfrac{I_T}{V} = \dfrac{2\text{ mA}}{48.6\text{ V}} = 41.2\ \mu\text{S}$, $R_T = \dfrac{1}{G_T} = 24.3\text{ k}\Omega$

$I_1 = \dfrac{V}{R_1} = \dfrac{48.6\text{ V}}{43\text{ k}\Omega} = 1.13\text{ mA}$, $I_2 = I_T - I_1 = 870\ \mu\text{A}$, $R_2 = \dfrac{V}{I_2} = 55.9\text{ k}\Omega$

7. a) $I_2 = \dfrac{V}{R_2} = \dfrac{12\text{ V}}{20\text{ k}\Omega} = 600\ \mu\text{A}$, $I_T = I_1 + I_2 = 1.13\text{ mA}$

$G_T = \dfrac{I_T}{V} = \dfrac{1.13\text{ mA}}{12\text{ V}} = 94.2\ \mu\text{S}$, $R_T = \dfrac{1}{G_T} = 10.6\text{ k}\Omega$, $R_1 = \dfrac{V}{I_1} = 22.6\text{ k}\Omega$

b) $I_2 = \dfrac{V}{R_2} = \dfrac{5\text{ V}}{15\text{ k}\Omega} = 333\ \mu\text{A}$, $I_T = I_1 + I_2 = 1.33\text{ mA}$

$G_T = \dfrac{I_T}{V} = \dfrac{1.33\text{ mA}}{5\text{ V}} = 267\ \mu\text{S}$, $R_T = \dfrac{1}{G_T} = 3.75\text{ k}\Omega$, $R_1 = \dfrac{V}{I_1} = 5\text{ k}\Omega$

c) $I_2 = \dfrac{V}{R_2} = \dfrac{30.9\text{ V}}{10\text{ k}\Omega} = 3.09\text{ mA}$, $I_T = I_1 + I_2 = 4.80\text{ mA}$

$G_T = \dfrac{I_T}{V} = \dfrac{4.80\text{ mA}}{30.9\text{ V}} = 155\ \mu\text{S}$, $R_T = \dfrac{1}{G_T} = 6.44\text{ k}\Omega$, $R_1 = \dfrac{V}{I_1} = 18.1\text{ k}\Omega$

d) $I_2 = \frac{V}{R_2} = \frac{2.16\ V}{91\ k\Omega} = 23.7\ \mu A,\ I_T = I_1 + I_2 = 287\ \mu A$

$G_T = \frac{I_T}{V} = \frac{287\ \mu A}{21.6\ V} = 133\ \mu S,\ R_T = \frac{1}{G_T} = 7.53\ k\Omega,\ R_1 = \frac{V}{I_1} = 8.21\ k\Omega$

9. a) $G_T = G_1 + G_2 + G_3 = 370\ \mu S + 813\ \mu S + 1\ mS = 2.2\ mS,\ R_T = \frac{1}{G_T} = 454\ \Omega$

$V = \frac{I_T}{G_T} = \frac{1.73\ mA}{2.2\ mS} = 785\ mV,\ I_1 = VG_1 = 291\ \mu A,\ I_2 = VG_2 = 654\ \mu A$

$I_3 = VG_3 = 785\ \mu A$

b) $G_T = G_1 + G_2 + G_3 = 123\ \mu S + 133\ \mu S + 213\ \mu S = 469\ \mu S,\ R_T = \frac{1}{G_T} = 2.13$

$V = \frac{I_T}{G_T} = \frac{500\ \mu A}{470\ \mu S} = 1.07\ V,\ I_1 = VG_1 = 131\ \mu A,\ I_2 = VG_2 = 142\ \mu A$

$I_3 = VG_3 = 227\ \mu A$

c) $G_T = G_1 + G_2 + G_3 = 45.4\ \mu S + 25.6\ \mu S + 17.9\ \mu S = 89\ \mu S$

$R_T = \frac{1}{G_T} = 11.2\ k\Omega$

$V = \frac{I_T}{G_T} = \frac{500\ \mu A}{89\ \mu S} = 5.62\ V,\ I_1 = VG_1 = 255\ \mu A,\ I_2 = VG_2 = 144\ \mu A$

$I_3 = VG_3 = 101\ \mu A$

d) $G_T = G_1 + G_2 + G_3 = 370\ \mu S + 256\ \mu S + 556\ \mu S = 1.18\ mS$

$R_T = \frac{1}{G_T} = 846\ \Omega$

$V = \frac{I_T}{G_T} = \frac{2\ mA}{1.18\ mS} = 1.69\ V,\ I_1 = VG_1 = 627\ \mu A,\ I_2 = VG_2 = 434\ \mu A$

$I_3 = VG_3 = 942\ \mu A$

11. a) $I_2 = \frac{V}{R_2} = \frac{12.6\ V}{12\ k\Omega} = 1.05\ mA,\ I_3 = I_T - (I_1+I_2) = 2.25\ mA$

$G_T = \frac{I_T}{V} = \frac{4\ mA}{12.6\ V} = 317\ \mu S,\ R_T = \frac{1}{G_T} = 3.15\ k\Omega$

$R_1 = \frac{V}{I_1} = 18\ k\Omega,\ R_3 = \frac{V}{I_3} = 5.6\ k\Omega$

b) $I_2 = \frac{V}{R_2} = \frac{20\ V}{270\ k\Omega} = 74.1\ \mu A,\ I_3 = I_T - (I_1+I_2) = 151\ \mu A$

$G_T = \frac{I_T}{V} = \frac{300\ mA}{20\ V} = 15\ \mu S,\ R_T = \frac{1}{G_T} = 66.7\ k\Omega$

$R_1 = \frac{V}{I_1} = 267\ k\Omega,\ R_3 = \frac{V}{I_3} = 133\ k\Omega$

c) $I_2 = \frac{V}{R_2} = \frac{23.4\ V}{750\ k\Omega} = 31.2\ \mu A,\ I_3 = I_T - (I_1+I_2) = 234\ \mu A$

$G_T = \frac{I_T}{V} = \frac{300\ \mu A}{23.4\ V} = 12.8\ \mu S,\ R_T = \frac{1}{G_T} = 78.0\ k\Omega$

$R_1 = \frac{V}{I_1} = 678\ k\Omega,\ R_3 = \frac{V}{I_3} = 99.9\ k\Omega$

d) $I_2 = \frac{V}{R_2} = \frac{13.8\ V}{680\ \Omega} = 20.3\ mA,\ I_3 = I_T - (I_1+I_2) = 24.6\ mA$

$G_T = \frac{I_T}{V} = \frac{60\ mA}{13.8\ V} = 4.35\ mS,\ R_T = \frac{1}{G_T} = 230\ \Omega$

$R_1 = \frac{V}{I_1} = 910\ \Omega,\ R_3 = \frac{V}{I_3} = 560\ \Omega$

13. a) $I_T = VG_T = 15\ V \times 1\ mS = 15\ mA,\ I_1 = VG_1 = 3.2\ mA$

$I_3 = I_T - (I_1+I_2) = 8.81\ mA,\ R_T = \frac{1}{G_T} = 1\ k\Omega,\ R_1 = \frac{V}{I_1} = 4.70\ k\Omega$

$R_2 = \frac{V}{I_2} = 5.00\ k\Omega,\ R_3 = \frac{V}{I_3} = 1.70\ k\Omega$

b) $I_T = VG_T = 30\ V \times 8\ mS = 240\ mA,\ I_1 = VG_1 = 60\ mA$

$I_3 = I_T - (I_1+I_2) = 120\ mA,\ R_T = \frac{1}{G_T} = 125\ \Omega,\ R_1 = \frac{V}{I_1} = 500\ \Omega$

$R_2 = \frac{V}{I_2} = 500\ \Omega,\ R_3 = \frac{V}{I_3} = 250\ \Omega$

c) $I_T = VG_T = 6.09\ V \times 3.29\ mS = 20\ mA,\ I_1 = VG_1 = 6.09\ mA$

$I_3 = I_T - (I_1+I_2) = 3.04\ mA,\ R_T = \frac{1}{G_T} = 304\ \Omega,\ R_1 = \frac{V}{I_1} = 1\ k\Omega$

$R_2 = \frac{V}{I_2} = 560\ \Omega,\ R_3 = \frac{V}{I_3} = 2\ k\Omega$

d) $I_T = VG_T = 5.29\ V \times 28.3\ \mu S = 150\ \mu A,\ I_1 = VG_1 = 70.6\ \mu A$

$I_3 = I_T - (I_1+I_2) = 26.5\ \mu A,\ R_T = \frac{1}{G_T} = 35.3\ k\Omega,\ R_1 = \frac{V}{I_1} = 75\ k\Omega$

$R_2 = \frac{V}{I_2} = 100\ k\Omega,\ R_3 = \frac{V}{I_3} = 200\ k\Omega$

***Problems 12-5***

1. a) $R_T = R_1 + R_2 \| R_3 = 10\ k\Omega + 19.4\ k\Omega = 29.4\ k\Omega$

$$I_T = \frac{E}{R_T} = \frac{15\ V}{29.4\ k\Omega} = 510\ \mu A,\ I_1 = I_T = 510\ \mu A$$

$V_1 = I_1R_1 = 510\ \mu A \times 10\ k\Omega = 5.1\ V,\ V_2 = V_3 = E - V_1 = 9.9\ V$

$$I_2 = \frac{V_2}{R_2} = \frac{9.9\ V}{33\ k\Omega} = 300\ \mu A,\ I_3 = \frac{V_3}{R_3} = \frac{9.9\ V}{47\ k\Omega} = 211\ \mu A$$

b) $R_T = R_1 + R_2 \| R_3 = 180\ k\Omega + 307\ \Omega = 487\ \Omega$

$$I_T = \frac{E}{R_T} = \frac{9\ V}{487\ \Omega} = 18.5\ mA,\ I_1 = I_T = 18.5\ mA$$

$V_1 = I_1R_1 = 18.5\ mA \times 180\ \Omega = 3.33\ V,\ V_2 = V_3 = E - V_1 = 5.67\ V$

$$I_2 = \frac{V_2}{R_2} = \frac{5.67\ V}{560\ \Omega} = 10.1\ mA,\ I_3 = \frac{V_3}{R_3} = \frac{5.67\ V}{680\ \Omega} = 8.34\ mA$$

3. a) $V_1 = I_1R_1 = 500\ \mu A \times 12\ k\Omega = 6\ V,\ V_2 = V_3 = E - V_1 = 4\ V$

$$I_2 = \frac{V_2}{R_2} = \frac{4\ V}{47\ k\Omega} = 85.1\ \mu A,\ I_3 = I_T - I_2 = 415\ \mu A$$

$$R_3 = \frac{V_3}{I_3} = \frac{4\ V}{415\ \mu A} = 9.64\ k\Omega,\ R_T = \frac{E}{I_T} = \frac{10\ V}{500\ \mu A} = 20\ k\Omega$$

b) $R_T = \frac{E}{I_T} = \frac{12\ V}{600\ \mu A} = 20\ k\Omega,\ R_1 = R_T - R_2 \| R_3 = 20\ k - 16\ k = 4.05\ k\Omega$

$V_1 = I_1R_1 = 600\ \mu A \times 4.05\ k\Omega = 2.43\ V,\ V_2 = V_3 = E - V_1 = 9.57\ V$

$$I_2 = \frac{V_2}{R_2} = \frac{9.57\ V}{39\ k\Omega} = 245\ \mu A,\ I_3 = \frac{V_2}{R_2} = \frac{9.57\ V}{27\ k\Omega} = 355\ \mu A$$

5. a) $R_T = R_3 \| (R_1+R_2) = 18\ k\Omega \| 34\ k\Omega = 11.8\ k\Omega,\ I_T = \frac{E}{R_T} = 5.10\ mA$

$$V_3 = E = 60\ V,\ I_3 = \frac{V_3}{R_3} = \frac{60\ V}{18\ k\Omega} = 3.33\ mA,\ I_1 = I_2 = I_T - I_3 = 1.77\ mA$$

$V_2 = I_2R_2 = 1.77\ mA \times 22\ k\Omega = 38.9\ V,\ V_1 = E - V_2 = 21.2\ V$

b) $R_T = R_3 \| (R_1+R_2) = 47\ k\Omega \| 34.5\ k\Omega = 19.9\ k\Omega,\ I_T = \frac{E}{R_T} = 151\ \mu A$

$$V_3 = E = 3\ V,\ I_3 = \frac{V_3}{R_3} = \frac{3\ V}{47\ k\Omega} = 63.8\ \mu A,\ I_1 = I_2 = I_T - I_3 = 87.2\ \mu A$$

$V_2 = I_2R_2 = 87.2\ \mu A \times 27\ k\Omega = 2.35\ V,\ V_1 = E - V_2 = 652\ mV$

7. a) $R_x = R_2 \| R_3 = 964\ \Omega$, $R_y = R_1 + R_x = 1.64\ k\Omega$, $R_T = R_y \| R_4 = 1.22\ k\Omega$

$I_T = \dfrac{E}{R_T} = 32.8\ mA$, $V_4 = E = 40\ V$, $I_4 = \dfrac{V_4}{R_4} = \dfrac{40\ V}{4.7\ k\Omega} = 8.51\ mA$

$I_1 = I_T - I_4 = 24.3\ mA$, $V_1 = I_1R_1 = 16.5\ V$, $V_2 = V_3 = E - V_1 = 23.5\ V$

$I_2 = \dfrac{V_2}{R_2} = \dfrac{23.5\ V}{1.5\ k\Omega} = 15.6\ mA$, $I_3 = \dfrac{V_3}{R_3} = \dfrac{23.5\ V}{2.7\ k\Omega} = 8.69\ mA$

b) $R_x = R_2 \| R_3 = 278\ \Omega$, $R_y = R_1 + R_x = 428\ k\Omega$, $R_T = R_y \| R_4 = 263\ k\Omega$

$I_T = \dfrac{E}{R_T} = \dfrac{30\ V}{263\ k\Omega} = 114\ \mu A$, $V_4 = E = 30\ V$, $I_4 = \dfrac{V_4}{R_4} = \dfrac{30\ V}{680\ k\Omega} = 44.1\ \mu A$

$I_1 = I_T - I_4 = 69.9\ \mu A$, $V_1 = I_1R_1 = 10.5\ V$, $V_2 = V_3 = E - V_1 = 19.5\ V$

$I_2 = \dfrac{V_2}{R_2} = \dfrac{19.5\ V}{470\ k\Omega} = 41.5\ \mu A$, $I_3 = \dfrac{V_3}{R_3} = \dfrac{19.5\ V}{680\ k\Omega} = 28.7\ \mu A$

9. a) $R_T = R_1 + R_2 \| R_3 \| R_4 = 3\ k\Omega + 5.96\ k\Omega = 8.96\ k\Omega$

$I_T = \dfrac{E}{R_T} = \dfrac{25\ V}{8.96\ k\Omega} = 2.79\ mA$, $I_1 = I_T = 2.79\ mA$

$V_1 = I_1R_1 = 8.37\ V$, $V_2 = V_3 = V_4 = E - V_1 = 16.6\ V$

$I_2 = \dfrac{V_2}{R_2} = 1.11\ mA$, $I_3 = \dfrac{V_3}{R_3} = 756\ \mu A$, $I_4 = \dfrac{V_4}{R_4} = 924\ \mu A$

b) $R_T = \dfrac{E}{I_T} = \dfrac{25\ V}{250\ \mu A} = 100\ k\Omega$, $V_3 = V_4 = V_2 = 10\ V$, $V_1 = E - V_2 = 15\ V$

$I_1 = I_T = 250\ \mu A$, $R_1 = \dfrac{V_1}{I_1} = \dfrac{15\ V}{250\ \mu A} = 60\ k\Omega$

$I_3 = \dfrac{V_3}{R_3} = \dfrac{10\ V}{150\ k\Omega} = 66.7\ \mu A$, $I_4 = \dfrac{V_4}{R_4} = \dfrac{10\ V}{200\ k\Omega} = 50\ \mu A$

$I_2 = I_T - (I_3+I_4) = 133\ \mu A$, $R_2 = \dfrac{V_2}{I_2} = \dfrac{10\ V}{133\ \mu A} = 75\ k\Omega$

11. a) $R_T = R_1 + R_2 \| R_3 \| R_4 = 10\ k\Omega + 33\ k\Omega + 18.2\ k\Omega = 61.2\ k\Omega$

$I_T = \dfrac{E}{R_T} = \dfrac{40\ V}{61.2\ k\Omega} = 653\ \mu A$, $I_1 = I_4 = I_T = 653\ \mu A$

$V_1 = I_1R_1 = 6.53\ V$, $V_4 = I_4R_4 = 21.6\ V$, $V_2 = V_3 = E - (V_1+V_4) = 11.9\ V$

$I_2 = \dfrac{V_2}{R_2} = 441\ \mu A$, $I_3 = \dfrac{V_3}{R_3} = 213\ \mu A$

b) $R_T = R_1 + R_2 \| R_3 \| R_4 = 150\ k\Omega + 150\ k\Omega + 278\ k\Omega = 578\ k\Omega$

$$I_T = \frac{E}{R_T} = \frac{15\ V}{578\ k\Omega} = 26\ \mu A,\ I_1 = I_4 = I_T = 26\ \mu A$$

$$V_1 = I_1 R_1 = 3.89\ V,\ V_4 = I_4 R_4 = 3.89\ V,\ V_2 = V_3 = E - (V_1 + V_4) = 7.21\ V$$

$$I_2 = \frac{V_2}{R_2} = \frac{7.21\ V}{470\ k\Omega} = 15.3\ \mu A,\ I_3 = \frac{V_3}{R_3} = \frac{7.21\ V}{680\ k\Omega} = 10.6\ \mu A$$

# CHAPTER 13

***Problems 13-1***

1. $I_2$ (with $E_2$ shorted) = 3.42 mA
   $I_2$ (with $E_1$ shorted) = 12.6 mA

   The currents are opposing and the resultant current is:
   $I_2$ = 12.6 mA - 3.42 mA = 9.13 mA

   $V_1$ = 41.4 V, $V_2$ = 16.4 V,

   $V_3$ = 33.6 V, $I_1$ = 18.8 mA,

   $I_2$ = 9.13 mA, $I_3$ = 28 mA

3. $I_3$ (with $E_2$ shorted) = 12.1 mA
   $I_3$ (with $E_1$ shorted) = 6.17 mA

   The currents are aiding and the resultant current is:
   $I_3$ = 12.1 mA + 6.17 mA = 18.2 mA

   $V_1$ = 36.8 V, $V_2$ = 45 V,

   $V_3$ = 18.2 V, $V_4$ = 21.8 V,

   $I_1$ = 13.6 mA, $I_2$ = 13.6 mA,

   $I_3$ = 18.2 mA, $I_4$ = 4.64 mA

5. $I_3$ (with $E_2$ shorted) = 1.01 mA
   $I_3$ (with $E_1$ shorted) = 1.42 mA

   The currents are aiding and the resultant current is:
   $I_3$ = 1.01 mA + 1.42 mA = 2.43 mA

   $V_1$ = 1 V, $V_2$ = 10 V, $V_3$ = 8 V,

   $I_1$ = 0.303 mA, $I_2$ = 2.13 mA,

   $I_3$ = 2.43 mA

7. $I_2$ (with $E_2$ shorted) = 17 mA
   $I_2$ (with $E_1$ shorted) = 4.63 mA

   The currents are aiding and the resultant current is:
   $I_2$ = 17 mA = 4.63 mA = 21.6 mA

   $V_1$ = 14.8 V, $V_2$ = 10.2 V,

   $V_3$ = 0.2 V, $I_1$ = 21.8 mA,

   $I_2$ = 21.6 mA, $I_3$ = 0.2 mA

9. $I_2$ (with $E_2$ shorted) = 3.96 mA
   $I_2$ (with $E_2$ shorted) = 1.53 mA

   The currents are aiding and the resultant current is:
   $I_3$ = 3.96 mA + 1.53 mA = 5.49 mA

   $V_1$ = 14.2 V, $V_2$ = 7.73 V,

   $V_3$ = 18.1 V, $V_4$ = 3.12 V,

   $I_1$ = 6.44 mA, $I_2$ = 6.44 mA,

   $I_3$ = 5.49 mA, $I_4$ = 945 $\mu$A

***Problems 13-2***

1. $$V_{OC} = \frac{ER_2}{R_1 + R_2} = \frac{30 \text{ V} \times 47 \text{ k}\Omega}{22 \text{ k}\Omega + 47 \text{ k}\Omega} = 20.4 \text{ V}, \; R_{TH} = R_1 || R_2 = 15 \text{ k}\Omega$$

   $$I_L = \frac{V_{OC}}{R_{TH} + R_L} = \frac{20.4 \text{ V}}{25 \text{ k}\Omega} = 8.16 \;\mu\text{A}$$

   $$V_L = I_L R_L = 816 \;\mu\text{A} \times 10 \text{ k}\Omega = 8.16 \text{ V}$$

3. $$V_{OC} = E = 12 \text{ V}, \; R_{TH} = R_1 + R_2 | R_3 = 680 \;\Omega + 320 \;\Omega = 1 \text{ k}\Omega$$

   $$I_L = \frac{V_{OC}}{R_{TH} + R_L} = \frac{12 \text{ V}}{1 \text{ k}\Omega + 750 \;\Omega} = 6.86 \text{ mA}$$

   $$V_L = I_L R_L = 6.86 \text{ mA} \times 750 \;\Omega = 5.14 \text{ V}$$

5. $V_{OC} = \dfrac{ER_3}{R_1 + R_2 + R_3} = \dfrac{50\ V \times 33\ k\Omega}{65\ k\Omega} = 25\ V$, $R_{TH} = R_3 \| (R_1+R_2) = 16.2\ k\Omega$

$$I_L = \frac{V_{OC}}{R_{TH} + R_L} = \frac{25.4\ V}{31.2\ k\Omega} = 814\ \mu A$$

$$V_L = I_L R_L = 814\ \mu A \times 15\ k\Omega = 12.2\ V$$

7. With $R_2$ open, $V_1 = 5.44$ V and $V_2 = 4.56$ V. Then $V_{OC} = 29.6$ V.

$$R_{TH} = R_2 \| R_3 = 2.56\ k\Omega$$

$$I_L = \frac{V_{OC}}{R_{TH} + R_L} = \frac{29.6\ V}{3.56\ k\Omega} = 8.3\ mA$$

$$V_L = I_L R_L = 8.3\ mA \times 1\ k\Omega = 8.3\ V$$

9. With $R_L$ open, $V_1 = 8.17$ V, $V_2 = 11.8$ V, $V_3 = 6.67$ V and $V_4 = 13.3$ V. Then $V_{OC} = 1.5$ V. $R_{TH} = R_1 \| R_2 + R_3 \| R_4 = 3.78\ k\Omega$

$$I_L = \frac{V_{OC}}{R_{TH} + R_L} = \frac{1.5\ V}{3.78\ k\Omega + 1\ k\Omega} = 314\ \mu A$$

$$V_L = I_L R_L = 314\ \mu A \times 1\ k\Omega = 314\ mV$$

11. $V_{OC} = \dfrac{V_{CC} R_2}{R_1 + R_2} = 2.53\ V$, $R_{TH} = R_1 \| R_2 = 5.94\ k\Omega$

$$R_L = R_4(H_{FE} + 1) = 470\ \Omega \times 81 = 38.1\ k\Omega$$

$$I_B = \frac{V_{OC} - V_{BE}}{R_{TH} + R_L} = 43.9\ \mu A,\ I_C = H_{FE} I_B = 80 \times 43.9\ \mu A = 3.51\ mA$$

$$V_C = V_{CC} - V_3 = 20\ V - 3.51\ mA \times 2\ k\Omega = 13\ V$$

***Problems 13-3***

1. a) $I_{SC} = \dfrac{E}{R_1} = \dfrac{30\ V}{22\ k\Omega} = 1.36\ mA$, $G_N = G_1 + G_2 = 66.7\ \mu S$

$$V_L = \frac{I_{SC}}{G_N + G_L} = \frac{1.36\ mA}{66.7\ \mu S + 100\ \mu S} = 8.16\ V$$

$$I_L = V_L G_L = 8.16\ V \times 100\ \mu S = 816\ \mu A$$

b) $I_{SC} = \dfrac{E}{R_1} = \dfrac{50\ V}{18\ k\Omega} = 2.78\ mA$, $G_N = G_1 + G_2 = 85.9\ \mu S$

$$V_L = \frac{I_{SC}}{G_N + G_L} = \frac{2.78\ mA}{85.9\ \mu S + 25.6\ \mu S} = 24.9\ V$$

$$I_L = V_L G_L = 24.9\ V \times 25.6\ \mu S = 639\ \mu A$$

3. a) $I_{SC} = \dfrac{E}{R_1 + R_2 \| R_3} = \dfrac{12\ \text{V}}{680\ \Omega + 320\ \Omega} = 12\ \text{mA}$, $R_{TH} = R_1 + R_2 \| R_3 = 1\ \text{k}\Omega$

$G_N = \dfrac{1}{1\ \text{k}\Omega} = 1\ \text{mS}$, $V_L = \dfrac{I_{SC}}{G_N + G_L} = \dfrac{12\ \text{mA}}{2.33\ \text{mS}} = 5.14\ \text{V}$

$I_L = V_L G_L = 5.14\ \text{V} \times 1.33\ \text{mS} = 6.86\ \text{mA}$

b) $I_{SC} = \dfrac{E}{R_1 + R_2 \| R_3} = \dfrac{10\ \text{V}}{12\ \text{k}\Omega + 8.18\ \text{k}\Omega} = 496\ \mu\text{A}$

$R_{TH} = R_1 + R_2 \| R_3 = 20.2\ \text{k}\Omega$

$G_N = \dfrac{1}{20.2\ \text{k}\Omega} = 49.6\ \mu\text{S}$

$V_L = \dfrac{I_{SC}}{G_N + G_L} = \dfrac{496\ \mu\text{A}}{49.6\ \mu\text{S} + 100\ \mu\text{S}} = 3.31\ \text{V}$

$I_L = V_L G_L = 331\ \mu\text{A}$

5. a) $I_{SC} = \dfrac{E}{R_1 + R_2} = 1.56\ \text{mA}$, $G_N = R_3 \| (R_1 + R_2) = 61.6\ \mu\text{S}$

$V_L = \dfrac{I_{SC}}{G_N + G_L} = \dfrac{1.56\ \text{mA}}{61.6\ \mu\text{S} + 66.7\ \mu\text{S}} = 12.2\ \text{V}$

$I_L = V_L G_L = 812\ \mu\text{A}$

b) $I_{SC} = \dfrac{E}{R_1 + R_2} = 1.52\ \text{mA}$, $G_N = R_3 \| (R_1 + R_2) = 299\ \mu\text{S}$

$V_L = \dfrac{I_{SC}}{G_N + G_L} = \dfrac{1.52\ \text{mA}}{299\ \mu\text{S} + 500\ \mu\text{S}} = 1.90\ \text{V}$

$I_L = V_L G_L = 1.9\ \text{V} \times 500\ \mu\text{S} = 951\ \mu\text{A}$

# CHAPTER 14

***Problems 14-1***

1\.

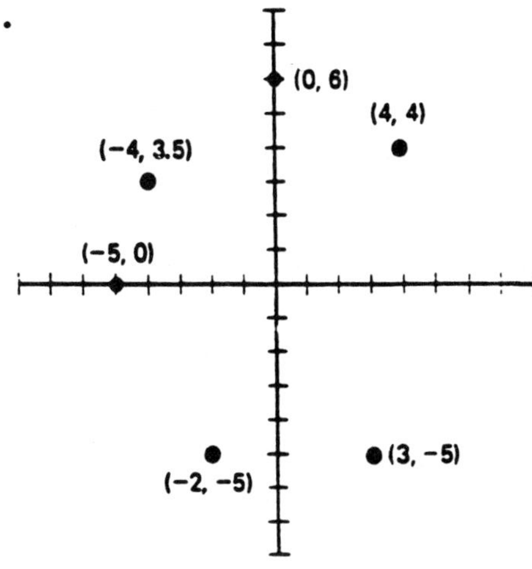


3\.

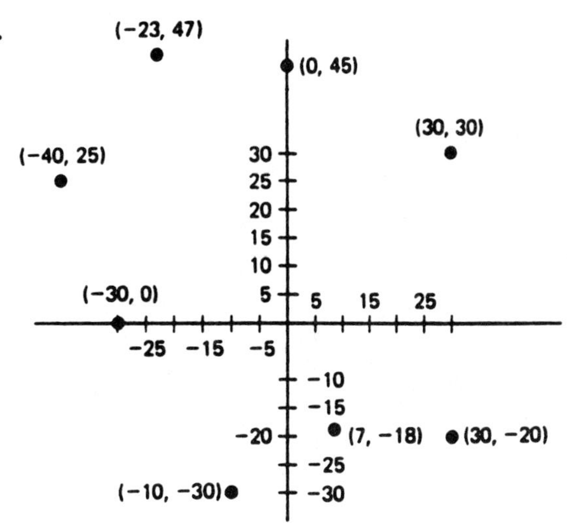


***Problems 14-2***

1\.

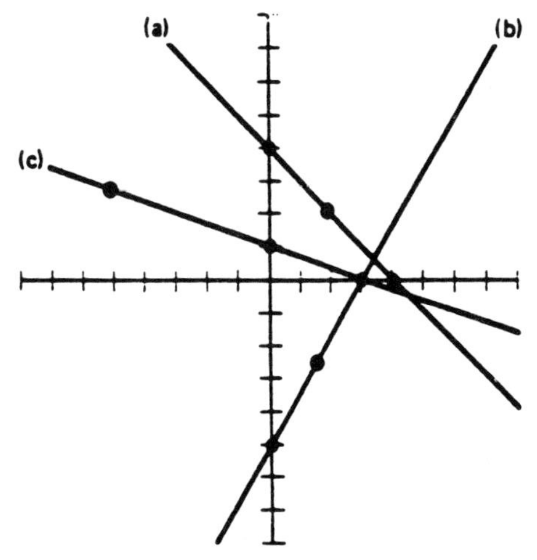




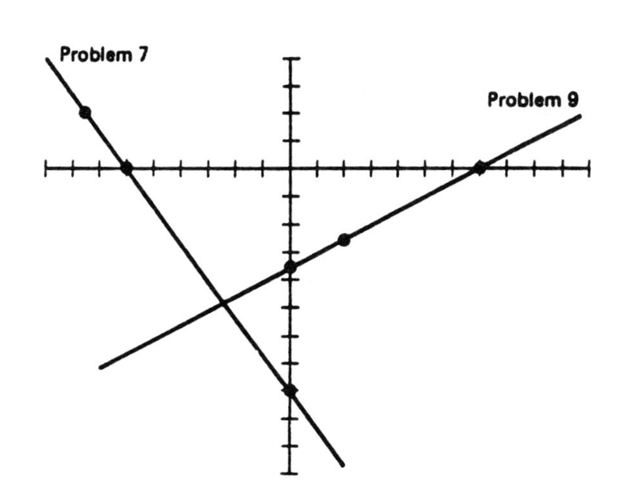

***Problems 14-3***

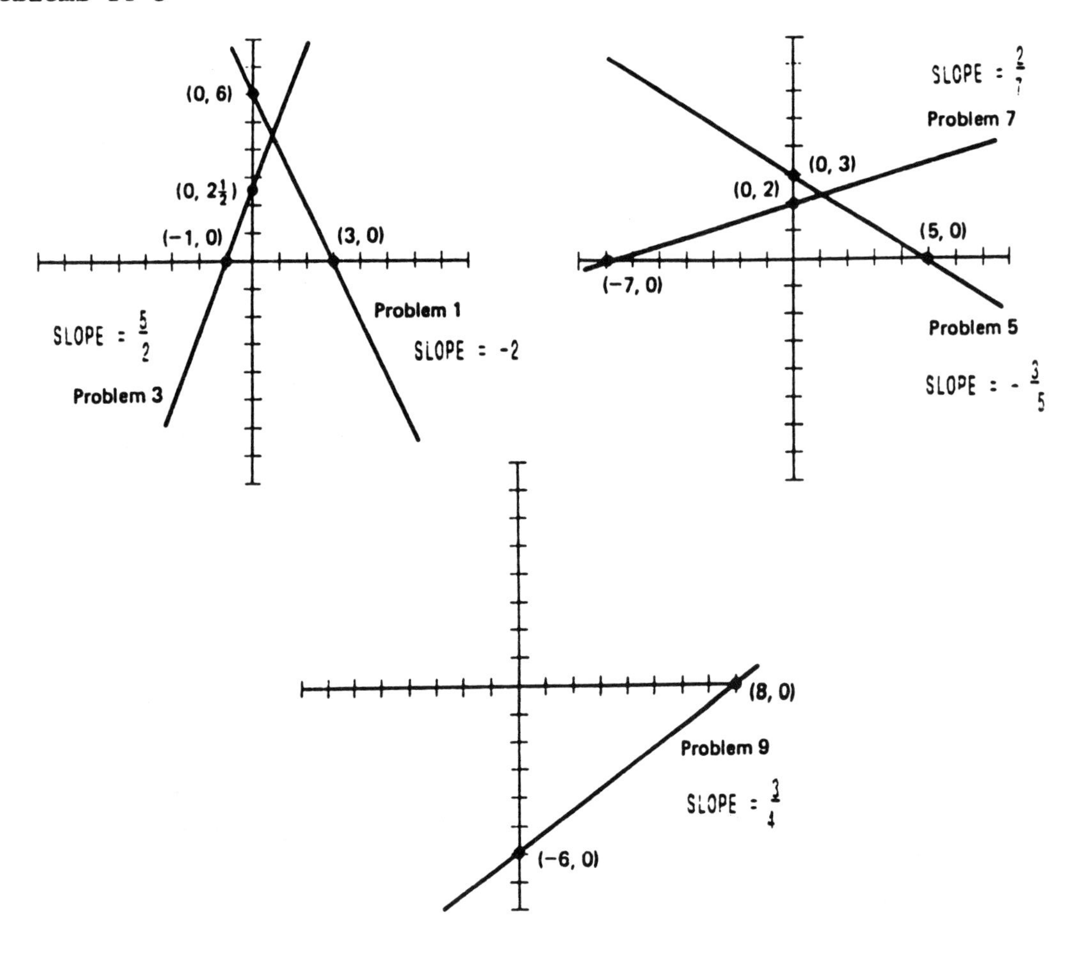


***Problems 14-4***

1. $Y = -\frac{1}{3}x + 1$, Slope $= -\frac{1}{3}$

   y-intercept = 0,1

3. $Y = -\frac{3}{4}x - \frac{5}{4}$; Slope $= -\frac{3}{4}$

   y-intercept = 0, $-\frac{5}{4}$

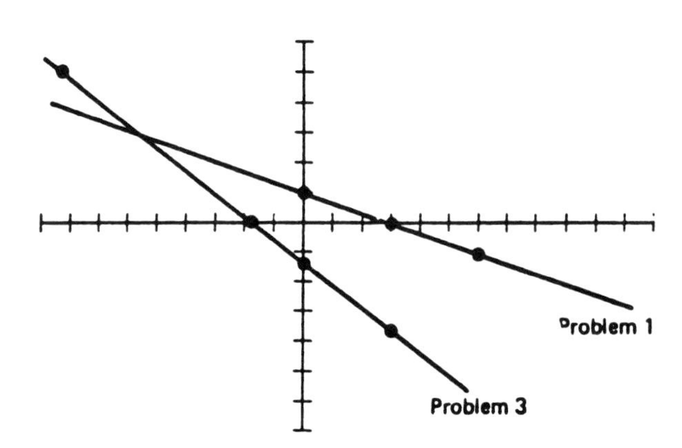

***Problems 14-4***

5. $Y = \frac{1}{4}x - 3$, Slope $= \frac{1}{4}$

   y-intercept = 0,-3

7. $Y = \frac{2}{5}x + 2$; Slope $= \frac{2}{5}$

   y-intercept = 0, 2

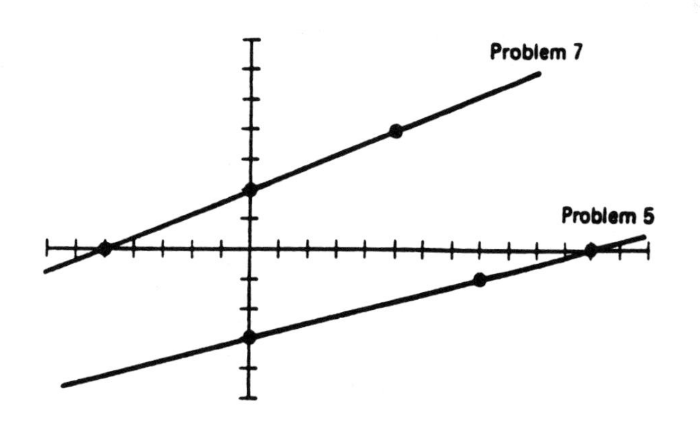


9. $Y = \frac{3}{7}x - 3$, Slope $= \frac{3}{7}$

   y-intercept = 0,-3

11. $Y = -\frac{1}{3}x + 6$; Slope $= -\frac{1}{3}$

   y-intercept = 0, 6

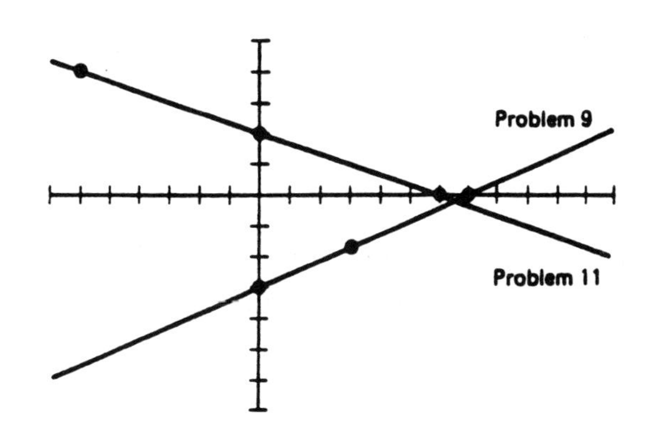


13. $Y = \frac{2}{3}x - \frac{8}{3}$; Slope $= \frac{2}{3}$

   y-intercept = 0, $-\frac{8}{3}$

15. $Y = \frac{1}{2}x - 3$, Slope $= \frac{1}{2}$

   y-intercept = 0,-3

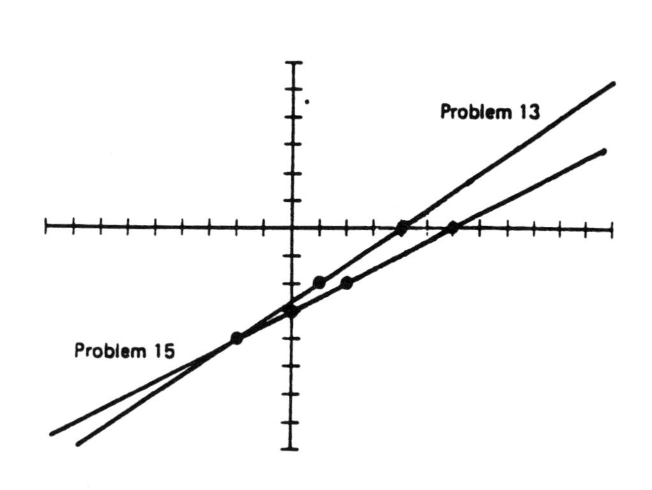

***Problems 14-4***

17. $Y = \frac{3}{2}x - 4$, Slope $= \frac{3}{2}$

    y-intercept = 0,-4

19. $Y = -3x - 9$; Slope = - 3

    y-intercept = 0, - 9

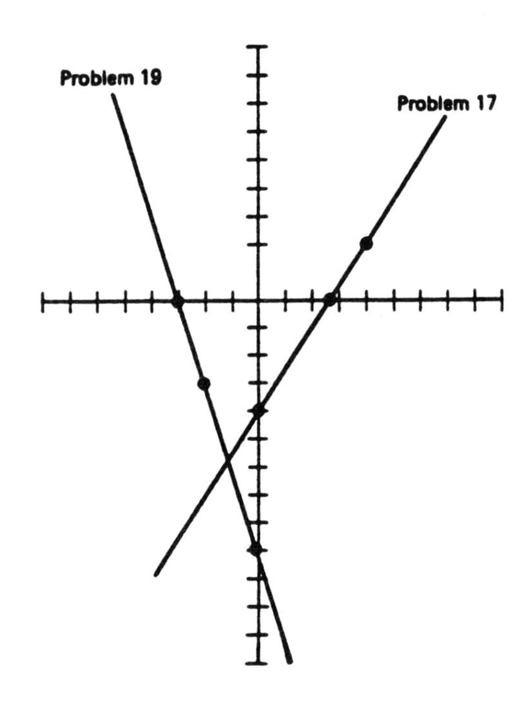


***Problems 14-5***

1. $R_{DC} \approx \frac{3\ V}{1.8\ mA} = 1.67\ k\Omega$. If V changes from 1 to 5 V ($\Delta V = 4$ V), then I changes from approximately 0.5 mA to approximately 4 mA ($\Delta I = 3.5$ mA).

   $r_{ac} = \frac{\Delta V}{\Delta I} \approx \frac{4\ V}{3.5\ mA} = 1.14\ k\Omega$

3. If $V_f = 0.4$ V, $I_f \approx 4.3$ mA; If $V_f = 0.6$ V, $I_f \approx 10.5$ mA.

   Then: $I_f \approx 6.2$ mA, Slope $= \frac{\Delta I}{\Delta V} = \frac{6.2\ mA}{0.2\ V} = 31$ mS and $r_d = 32.3\ \Omega$

5. (a) $V_C \approx 40\%$ of 100 V = 40 V, $V_R \approx 60\%$ of 100 V = 60 V
   (b) $V_C \approx 95\%$ of 100 V = 95 V, $V_R \approx 5\%$ of 100 V = 5 V

7. (a) $V_C \approx 78\%$ of 20 V = 15.6 V, $V_R \approx 22\%$ of 20 V = 4.4 V
   (b) $V_C \approx 98\%$ of 20 V = 19.6 V, $V_R \approx 2\%$ of 20 V = 0.4 V

***Problems 14-6***

1.

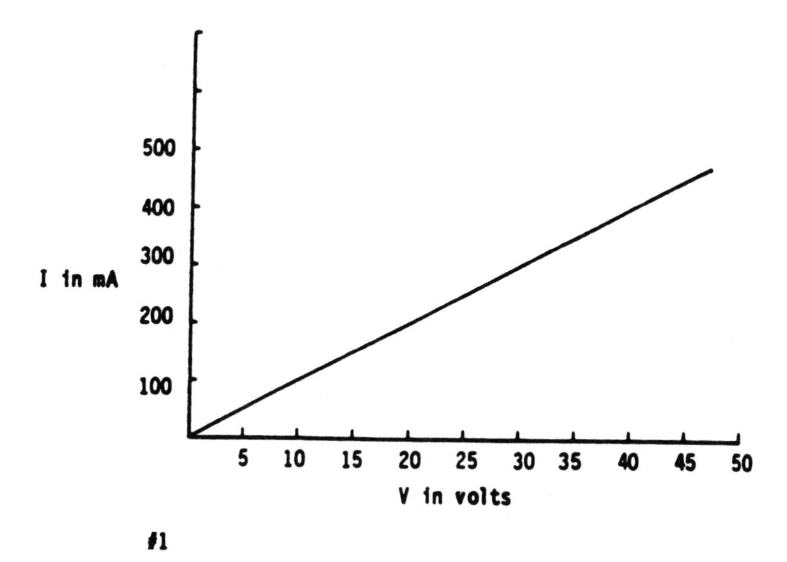


#1

3.

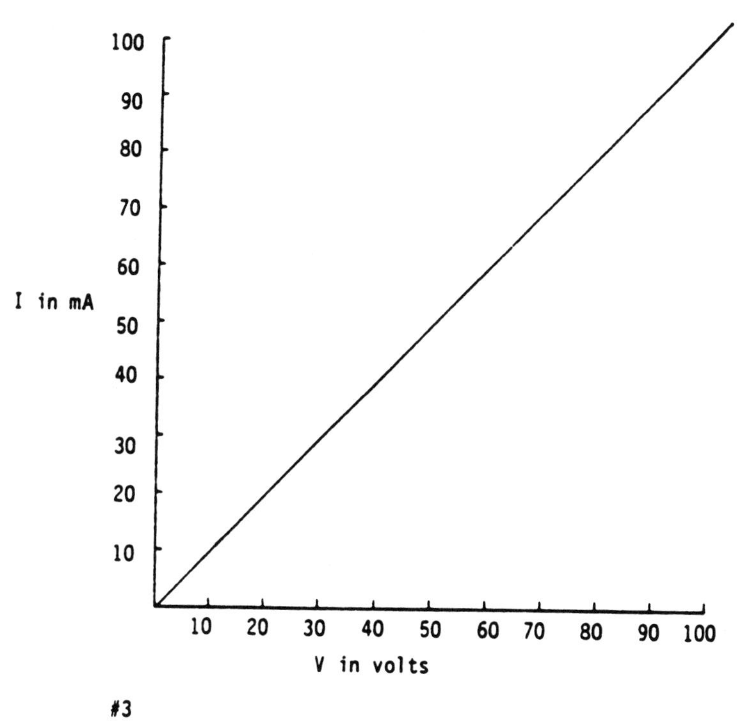


#3

***Problems 14-6***

5.

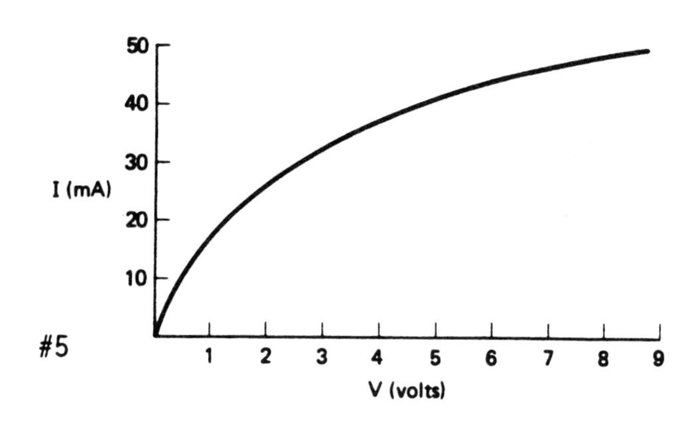


#5

7.

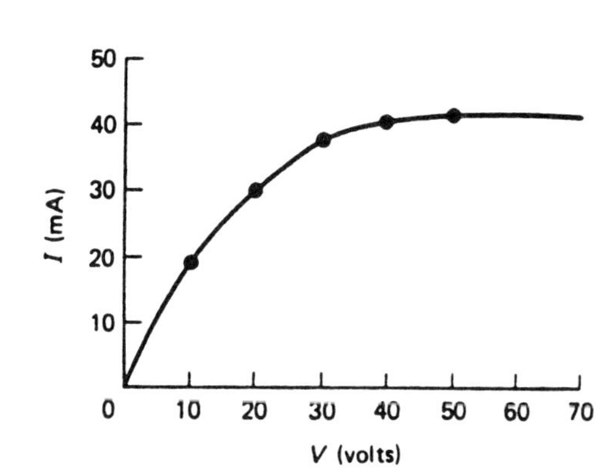


#7

# CHAPTER 15

***Problems 15-1***

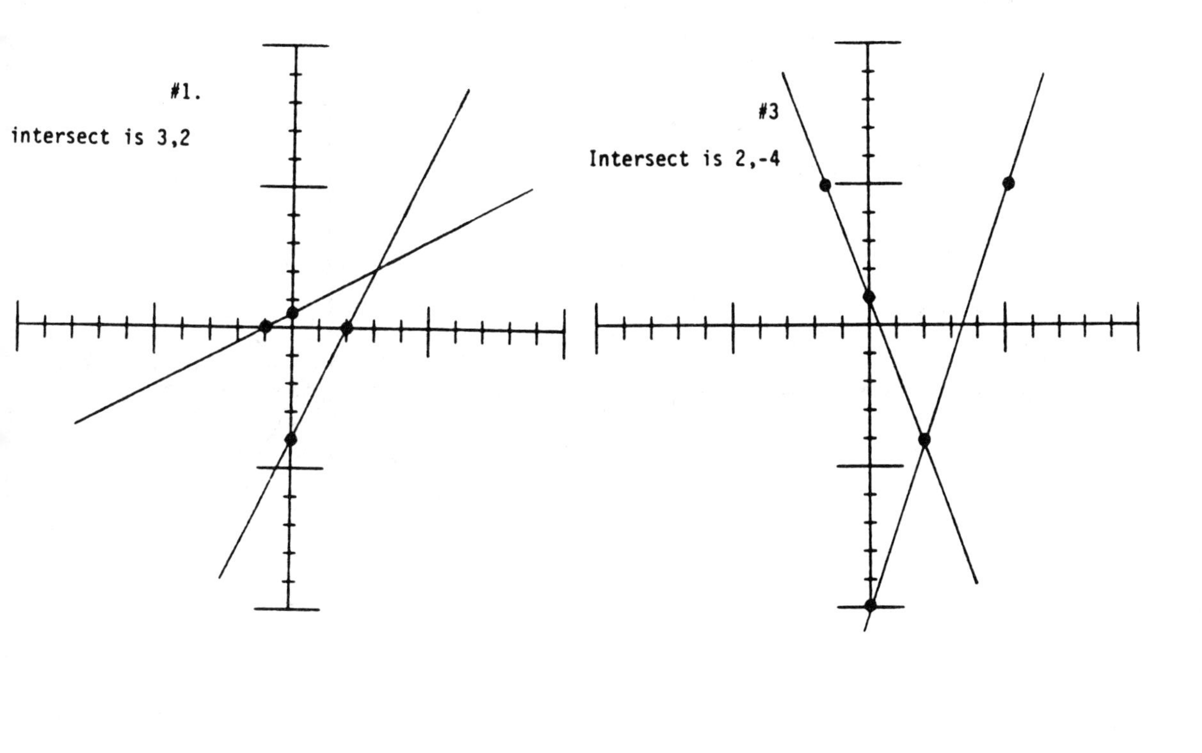


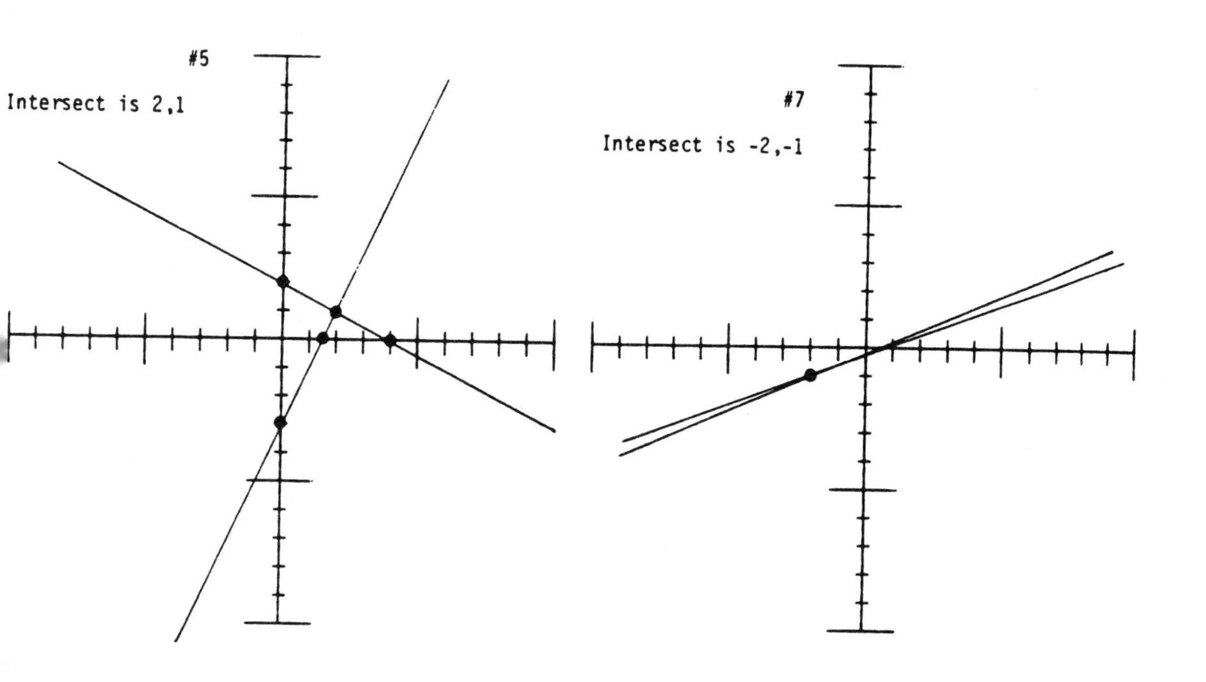

***Problems 15-1***

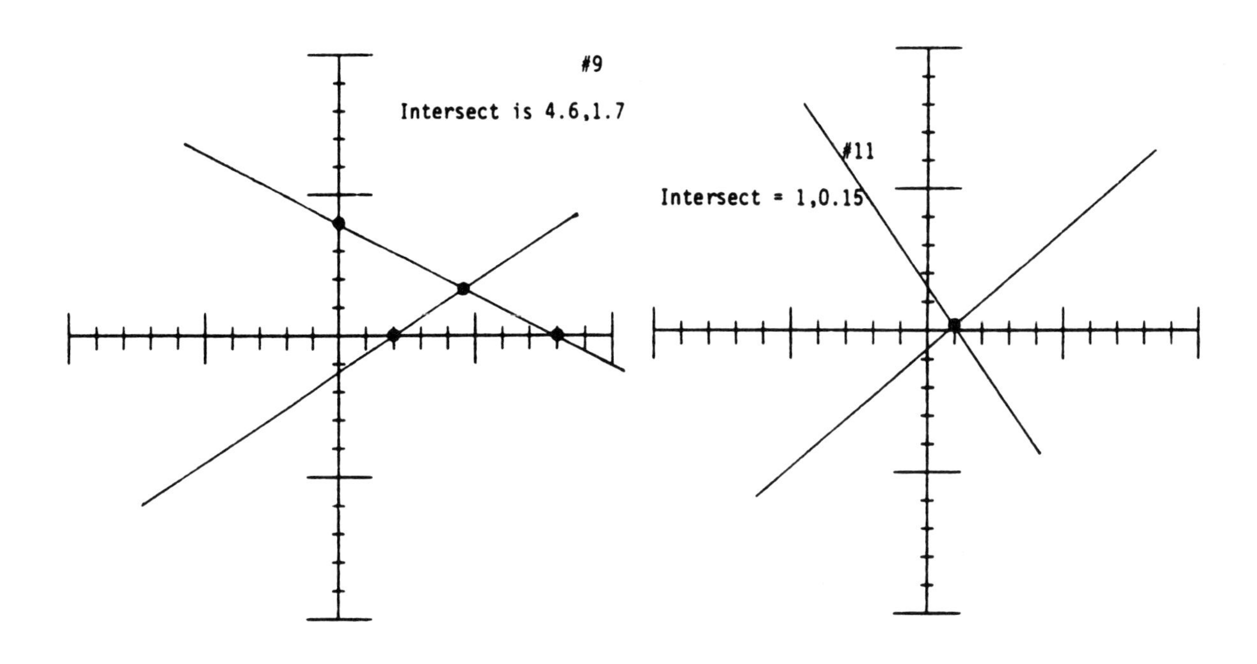


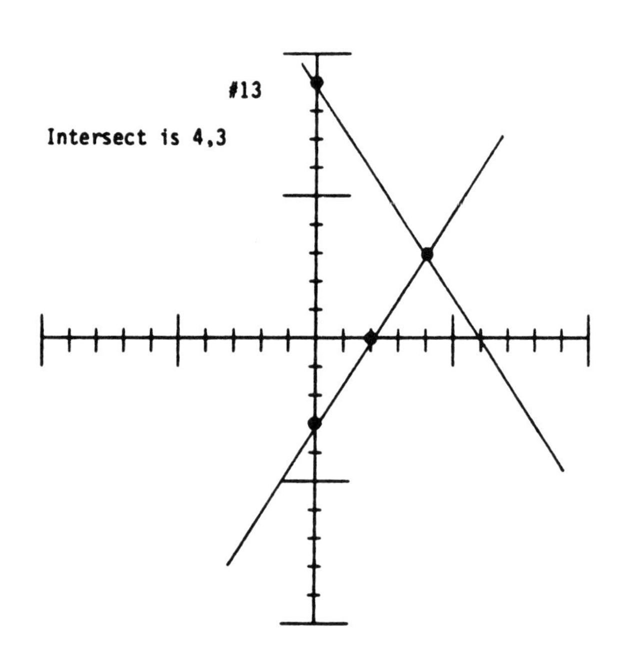

***Problems 15-2***

1. $2x - y = 4$ ; $2y = x + 1$

$2x - y = 4$
$2x - 4y = -2$ M x 2
$3y = 6$ subtr
$y = 2$

$2x - (2) = 4$
$2x = 6$
$x = 3$

coordinates are 3, 2

3. $5x + 2y = 2$ ; $3x - y = 10$

$5x + 2y = 2$
$6x - 2y = 20$ M x 2
$11x + 0 = 22$ add
$x = 2$

$5(2) + 2y = 4$
$2y = -8$
$y = -4$

coordinates are 2, -4

5. $8x - 4y = 12$ ; $x + 2y = 4$

$8x - 4y = 12$
$2x + 4y = 8$ M x 2
$10x + 0 = 20$ add
$x = 2$

$8(2) - 4y = 12$
$-4y = -4$
$y = 1$

coordinates are 2, 1

7. $2x - 5y = 1$ ; $3x - 8y = 2$

$6x - 15y = 3$ M x 3
$6x - 16y = 4$ M x 4
$y = -1$ subtr

$2x - 5(-1) = 1$
$3x = -4$
$x = -2$

coordinates are -2, -1

9. $\frac{x}{2} + \frac{y}{3} = 3$

$\frac{x}{2} - \frac{y}{3} = 1$

$\frac{2x}{2} = 4$ add

$x = 4$

$\frac{(4)}{2} + \frac{y}{3} = 3$

$\frac{y}{3} = 3 - 2$

$y = 3$

coordinates are 4, 3

11. $2x - 3y = 4$ ; $x + 2y = 8$

$2x - 3y = 4$
$2x + 4y = 16$ M x 2
$-7y = -12$
$y = 1.71$

$2x - 3(1.71) = 4$
$x = 4.57$

coordinates are 4.57, 1.71

13. $\frac{x}{3} + \frac{y}{2} = \frac{5}{12}$ ; $\frac{x}{4} - \frac{y}{3} = \frac{5}{24}$

$4x + 6y = 5$ M x 12

$4x - \frac{16y}{3} = \frac{10}{3}$ M x 16

$\frac{34y}{3} = \frac{5}{3}$ subtr

$y = 0.147$

$4x + 6(0.147) = 5$
$4x = 5 - 6(0.147)$
$x = 1.03$

coordinates are 1.03, 0.147

***Problems 15-3***

1. $x + 2y = 4 \quad (1)$
   $2x - y = 2 \quad (2)$

   solve for x in (1): $x = 4 - 2y$

   substitute for x in (2):
   $$2(4-2y) - y = 2$$
   $$8 - 4y - y = 2$$
   $$-5y = -6$$
   $$y = 1.2$$
   $$x + 2(1.2) = 4$$
   $$x = 1.6$$

   $x = 1.6$ and $y = 1.2$

3. $4x - 2y = 3 \quad (1)$
   $3x + 2y = 5 \quad (2)$

   solve for x in (1):
   $$x = \frac{3 + 2y}{4}$$

   substitute for x in (2):
   $$\frac{3(3+2y)}{4} + 2y = 5$$
   $$y = 0.786$$
   $$4x - 2(0.786) = 3$$
   $$4x = 3 + 2(0.786)$$
   $$x = 1.14$$

   $x = 1.14$ and $y = 0.786$

5. $5x + 3y = 5 \quad (1)$
   $3x - 4y = 6 \quad (2)$

   solve for y in (1):
   $$y = \frac{5 - 5x}{3}$$

   substitute for y in (2):
   $$3x - \frac{4(5-5x)}{3} = 6$$
   $$x = 1.31$$
   $$5(1.31) + 3y = 5$$
   $$3y = 5 - 5(1.31)$$
   $$y = -0.517$$

   $x = 1.31$ and $y$ $-0.517$

7. $x + y = 10 \quad (1)$
   $4x - 4y = 0 \quad (2)$

   solve for x in (1): $x = 10 - y$

   substitute for x in (2):
   $$4(10-y) - 4y = 0$$
   $$y = 5$$
   $$x + 5 = 10$$
   $$x = 5$$

   $x = 5$ and $y = 5$

9. $3x - 3y = -5 \quad (1)$
   $-x - y = 4 \quad (2)$

   solve for x in (1):
   $$x = \frac{3y - 5}{3}$$

   substitute for x in (2):
   $$\frac{-(3y-5)}{3} - y = 4$$
   $$3x - 3(-1.17) = -5$$
   $$3x = -5 + 3(-1.17)$$
   $$x = -2.83$$

   $x = -2.83$ and $y = -1.17$

***Problems 15-4***

1. $x + 2y = 4$
   $2x - y = 2$

$$x = \frac{\begin{vmatrix} 4 & 2 \\ 2 & -1 \end{vmatrix}}{\begin{vmatrix} 1 & 2 \\ 2 & -1 \end{vmatrix}} = \frac{4(-1) - 2(2)}{1(-1) - 2(2)} = 1.6$$

$$y = \frac{\begin{vmatrix} 1 & 4 \\ 2 & 2 \end{vmatrix}}{\begin{vmatrix} 1 & 2 \\ 2 & -1 \end{vmatrix}} = \frac{1(2) - 2(4)}{1(-1) - 2(2)} = 1.2$$

3. $4x - 2y = 3$
$3x + 2y = 5$

$$x = \frac{\begin{vmatrix} 3 & -2 \\ 5 & 2 \end{vmatrix}}{\begin{vmatrix} 4 & -2 \\ 3 & 2 \end{vmatrix}} = \frac{3(2) - 5(-2)}{4(2) - 3(-2)} = 1.14$$

$$y = \frac{\begin{vmatrix} 4 & 3 \\ 3 & 5 \end{vmatrix}}{\begin{vmatrix} 4 & -2 \\ 3 & 2 \end{vmatrix}} = \frac{4(5) - 3(3)}{4(2) - 3(-2)} = 0.786$$

5. $5x + 3y = 5$
$3x - 4y = 6$

$$x = \frac{\begin{vmatrix} 5 & 3 \\ 6 & -4 \end{vmatrix}}{\begin{vmatrix} 5 & 3 \\ 3 & -4 \end{vmatrix}} = \frac{5(-4) - 6(3)}{5(-4) - 3(3)} = 1.31$$

$$y = \frac{\begin{vmatrix} 5 & 5 \\ 3 & 6 \end{vmatrix}}{\begin{vmatrix} 5 & 3 \\ 3 & -4 \end{vmatrix}} = \frac{5(6) - 3(5)}{5(-4) - 3(3)} = -0.517$$

7. $x + y = 10$
$4x - 4y = 0$

$$x = \frac{\begin{vmatrix} 10 & 1 \\ 0 & -4 \end{vmatrix}}{\begin{vmatrix} 1 & 1 \\ 4 & -4 \end{vmatrix}} = \frac{10(-4) - 0(1)}{1(-4) - 1(4)} = 5$$

$$y = \frac{\begin{vmatrix} 1 & 10 \\ 4 & 0 \end{vmatrix}}{\begin{vmatrix} 1 & 1 \\ 4 & -4 \end{vmatrix}} = \frac{1(0) - 4(10)}{1(-4) - 1(4)} = 5$$

9. $3x - 3y = -5$
$-x - y = 4$

$$x = \frac{\begin{vmatrix} -5 & -3 \\ 4 & -1 \end{vmatrix}}{\begin{vmatrix} 3 & -3 \\ -1 & -1 \end{vmatrix}} = \frac{-5(-1) - 4(-3)}{3(-1) - (-1)(-3)} = -2.83$$

$$y = \frac{\begin{vmatrix} 3 & -5 \\ -1 & 4 \end{vmatrix}}{\begin{vmatrix} 3 & -3 \\ -1 & -1 \end{vmatrix}} = \frac{3(4) - (-1)(-5)}{3(-1) - (-1)(-3)} = -1.17$$

***Problems 15-5***

1. 
$$\begin{aligned} x + 2y + z &= 4 \\ 2x - y + 3z &= 3 \\ 2x - 2y + 3z &= 2 \end{aligned}$$

$$x = \frac{\begin{vmatrix} 4 & 2 & 1 \\ 3 & -1 & 3 \\ 2 & -2 & 3 \end{vmatrix}}{\begin{vmatrix} 1 & 2 & 1 \\ 2 & -1 & 3 \\ 2 & -2 & 3 \end{vmatrix}} = \frac{2}{1} = 2$$

$$y = \frac{\begin{vmatrix} 1 & 4 & 1 \\ 2 & 3 & 3 \\ 2 & 2 & 3 \end{vmatrix}}{\begin{vmatrix} 1 & 2 & 1 \\ 2 & -1 & 3 \\ 2 & -2 & 3 \end{vmatrix}} = \frac{1}{1} = 1 \qquad z = \frac{\begin{vmatrix} 1 & 2 & 4 \\ 2 & -1 & 3 \\ 2 & -2 & 2 \end{vmatrix}}{\begin{vmatrix} 1 & 2 & 1 \\ 2 & -1 & 3 \\ 2 & -2 & 3 \end{vmatrix}} = \frac{0}{1} = 0$$

3. 
$$\begin{aligned} x - 2y + 4z &= 4 \\ x + 3y - 2z &= 6 \\ 3x + y + 3z &= 5 \end{aligned}$$

$$x = \frac{\begin{vmatrix} 4 & -2 & 4 \\ 6 & 3 & -2 \\ 5 & 1 & 3 \end{vmatrix}}{\begin{vmatrix} 1 & -2 & 4 \\ 1 & 3 & -2 \\ 3 & 1 & 3 \end{vmatrix}} = \frac{64}{-3} = -21.3$$

$$y = \frac{\begin{vmatrix} 1 & 4 & 4 \\ 1 & 6 & -2 \\ 3 & 5 & 3 \end{vmatrix}}{\begin{vmatrix} 1 & -2 & 4 \\ 1 & 3 & -2 \\ 3 & 1 & 3 \end{vmatrix}} = \frac{-60}{-3} = 20 \qquad z = \frac{\begin{vmatrix} 1 & -2 & 4 \\ 1 & 3 & 6 \\ 3 & 1 & 5 \end{vmatrix}}{\begin{vmatrix} 1 & -2 & 4 \\ 1 & 3 & -2 \\ 3 & 1 & 3 \end{vmatrix}} = \frac{-49}{-3} = 16.3$$

5. 
$$\begin{aligned} x + y + z &= 12 \\ 2x - 3y - 4z &= 8 \\ -3x + 2y + 2z &= 5 \end{aligned}$$

$$x = \frac{\begin{vmatrix} 12 & -1 & 1 \\ 8 & 3 & -4 \\ 5 & 2 & 2 \end{vmatrix}}{\begin{vmatrix} 1 & 1 & 1 \\ 2 & -3 & -4 \\ -3 & 2 & 2 \end{vmatrix}} = \frac{19}{5} = 3.8$$

$$y = \frac{\begin{vmatrix} 1 & 12 & 1 \\ 2 & 8 & -4 \\ -3 & 5 & 2 \end{vmatrix}}{\begin{vmatrix} 1 & 1 & 1 \\ 2 & -3 & -4 \\ -3 & 2 & 2 \end{vmatrix}} = \frac{166}{5} = 33.2 \qquad z = \frac{\begin{vmatrix} 1 & 1 & 12 \\ 2 & -3 & 8 \\ -3 & 2 & 5 \end{vmatrix}}{\begin{vmatrix} 1 & 1 & 1 \\ 2 & -3 & -4 \\ -3 & 2 & 2 \end{vmatrix}} = \frac{125}{5} = -25$$

***Problems 15-6***

1. $20 - I_1R_1 + I_2R_2 = R_1(I_2 + I_3) + R_2I_2 = 4.7\text{ k } I_2 + 4.7\text{ k } I_3 + 10\text{ k } I_2$

   $20 - I_1R_1 + I_3R_3 = R_1(I_2 + I_3) + R_3I_3 = 4.7\text{ k } I_2 + 4.7\text{ k } I_3 + 20\text{ k } I_3$

   (1) $20 = 14.7\text{ k } I_2 + 4.7\text{ k } I_3$

   (2) $20 = 4.7\text{ k } I_2 + 24.7\text{ k } I_3$

   $I_1 = 1.76$ mA, $I_2 = 1.17$ mA, $I_3 = 587\ \mu$A, $V_1 = 8.27$ V, $V_2 = V_3 = 11.7$ V

***Problems 15-7***

*1. Starting from the positive side of $E_1$ and moving clockwise:*

   (1) $I_1R_1 + I_2R_2 - 20\text{ V} = 0$

   $20 = I_1R_1 + I_2R_2$

   $20 = R_1(I_2+I_3) + I_2R_2$

   $20 = 3\text{ k } I_2 + 3\text{ k } I_3 + 2\text{ k } I_2$

   $20 = 5\text{ k } I_2 + 3\text{ k } I_3$

   (2) $I_1R_1 - 40\text{ V} + I_3R_3 - 20\text{ V} = 0$

   $60 = I_1R_1 + I_3R_3$

   $60 = R_1(I_2+I_3) + I_3R_3$

   $60 = 3\text{ k } I_2 + 3\text{ k } I_3 + 1\text{ k } I_3$

   $60 = 3\text{ k } I_2 + 4\text{ k } I_3$

   $I_1 = 12.7$ mA, $I_2 = 9.08$ mA, $I_3 = 21.8$ mA, $V_1 = 38.2$ V, $V_2 = 18.2$ V, $V_3 = 21.8$ V

*3. Starting from the positive side of $E_1$ and moving clockwise:*

   (1) $I_1R_1 + I_2R_2 + 40\text{ V} - 100\text{ V} = 0$

   $60 = I_1R_1 + I_2R_2$

   $60 = R_1(I_2+I_3) + I_2R_2$

   $60 = 12\text{ k } I_2 + 12\text{ k } I_3 + 20\text{ k } I_2$

   $60 = 32\text{ k } I_2 + 12\text{ k } I_3$

   (2) $I_1R_1 + I_3R_3 - 100\text{ V} = 0$

   $100 = I_1R_1 + I_3R_3$

   $100 = R_1(I_2+I_3) + I_3R_3$

   $100 = 12\text{ k } I_2 + 12\text{ k } I_3 + 10\text{ k } I_3$

   $100 = 12\text{ k } I_2 + 22\text{ k } I_3$

   $I_1 = 4.63$ mA, $I_2 = 220\ \mu$A, $I_3 = 4.44$ mA, $V_1 = 55.6$ V, $V_2 = 4.44$ V, $V_3 = 44.4$ V

   (3) $I_1R_1 + I_3R_3 - 25\text{ V} = 0$

   $25 = I_1R_1 + I_3R_3$

   $25 = R_1(I_2+I_3) + I_3R_3$

   $25 = 470\ I_2 + 470\ I_3 + 330\ I_3$

   $25 = 470\ I_2 + 800\ I_3$

   $I_1 = 11.1$ mA, $I_2 = 49$ mA, $I_3 = 60$ mA, $V_1 = 5.2$ V, $V_2 = 40.2$ V, $V_3 = 19.8$ V

5. *Starting from the positive side of $E_1$ and moving counterclockwise:*

(1) $I_3R_3 + 30\text{ V} + I_1R_1 - 40\text{ V} = 0$

$10 = I_1R_1 + I_3R_3$

$10 = R_1(I_2+I_3) + R_3I_3$

$10 = 100\ I_2 + 100\ I_3 + 220\ I_3$

$10 = 100\ I_2 + 330\ I_3$

(2) $I_2R_2 + I_1R_1 - 40\text{ V} = 0$

$40 = I_1R_1 + I_2R_2$

$40 = R_1(I_2+I_3) + R_2I_2$

$40 = 100\ I_2 + 100\ I_3 + 330\ I_2$

$40 = 430\ I_2 + 100\ I_3$

$I_1 = 251\text{ mA}$, $I_2 = 45.3\text{ mA}$, $I_3 = 204\text{ mA}$, $V_1 = 25.1\text{ V}$, $V_2 = 14.9\text{ V}$, $V_3 = 44.9\text{ V}$

7. *Starting from the positive side of $E_1$ and moving counterclockwise:*

(1) $I_2R_2 - 8\text{ V} + I_1R_1 - 12\text{ V} = 0$

$20 = I_1R_1 + I_2R_2$

$20 = R_1(I_2+I_3) + I_2R_2$

$20 = 2.7\text{ k}\ I_2 + 2.7\text{ k}\ I_3 + 3.3\text{ k}\ I_2$

$20 = 6\text{ k}\ I_2 + 2.7\text{ k}\ I_3$

(2) $I_3R_3 + I_1R_1 - 12\text{ V} = 0$

$12 = I_1R_1 + I_3R_3$

$12 = R_1(I_2+I_3) + I_3R_3$

$12 = 2.7\text{ k}\ I_2 + 2.7\text{ k}\ I_3 + 1\text{ k}\ I_3$

$12 = 2.7\text{ k}\ I_2 + 3.7\text{ k}\ I_3$

$I_1 = 4\text{ mA}$, $I_2 = 2.79\text{ mA}$, $I_3 = 1.2\text{ mA}$, $V_1 = 10.8\text{ V}$, $V_2 = 9.2\text{ V}$, $V_3 = 1.2\text{ V}$

# CHAPTER 16

***Problems 16-1***

| | | | |
|---|---|---|---|
| 1. | j13 | 3. | j |
| 5. | -25 | 7. | 18 |
| 9. | j35 | 11. | j6.60 |
| 13. | -j18 | 15. | 12 |
| 17. | j3 | 19. | -j2 |
| 21. | 7 | 23. | -j20 |

***Problems 16-2***

| | | | |
|---|---|---|---|
| 1. | -4 - j4 | 3. | 4 - j8 |
| 5. | 3.1 - j4.5 | 7. | 11 + j7 |
| 9. | 1.9 - j4.2 | 11. | -1 - j7 |
| 13. | 10 - j6 | 15. | 0 +j0 |
| 17. | 6 + j0 | 19. | 0 - j6 |

***Problems 16-2***

**21, 22, 23, 24.**

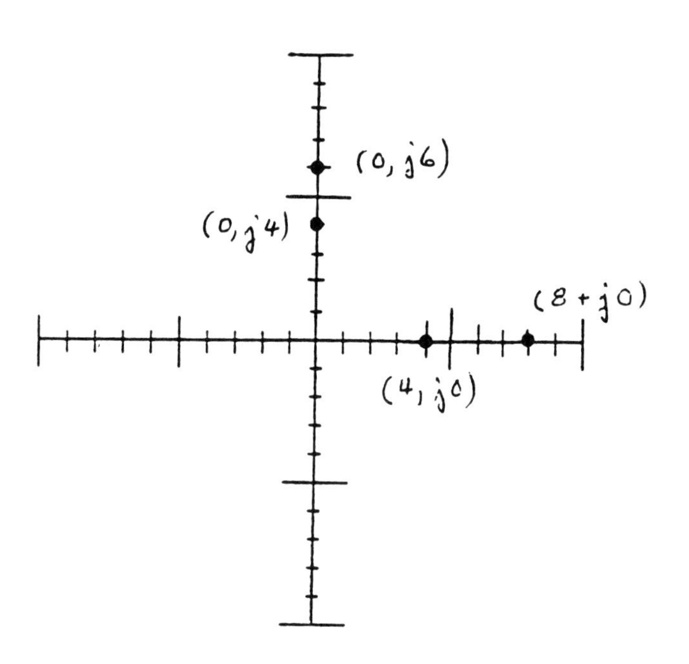


**25, 26, 27.**

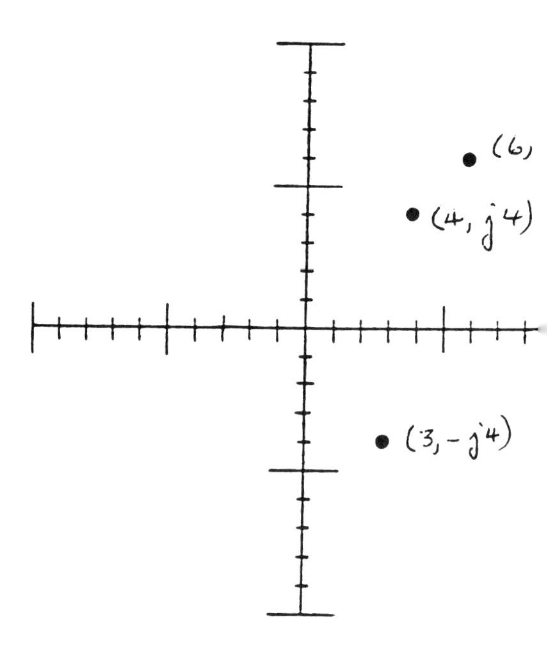


**28, 29, 30.**

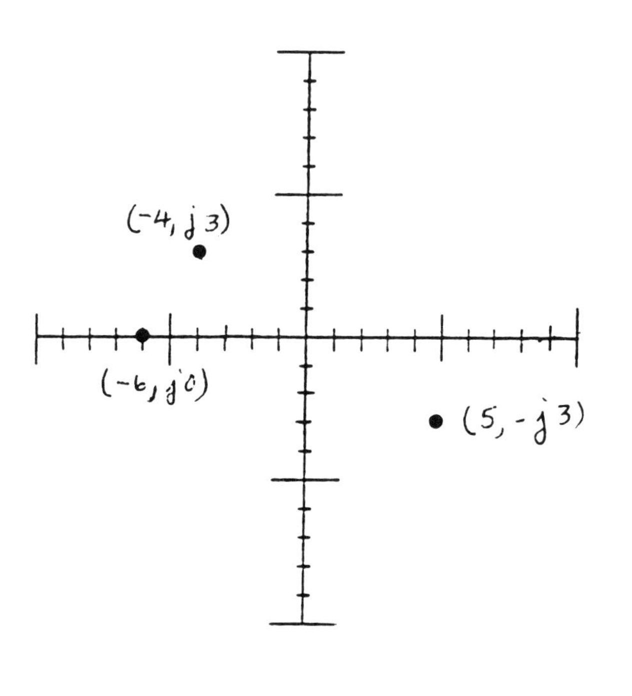


**31, 32.**

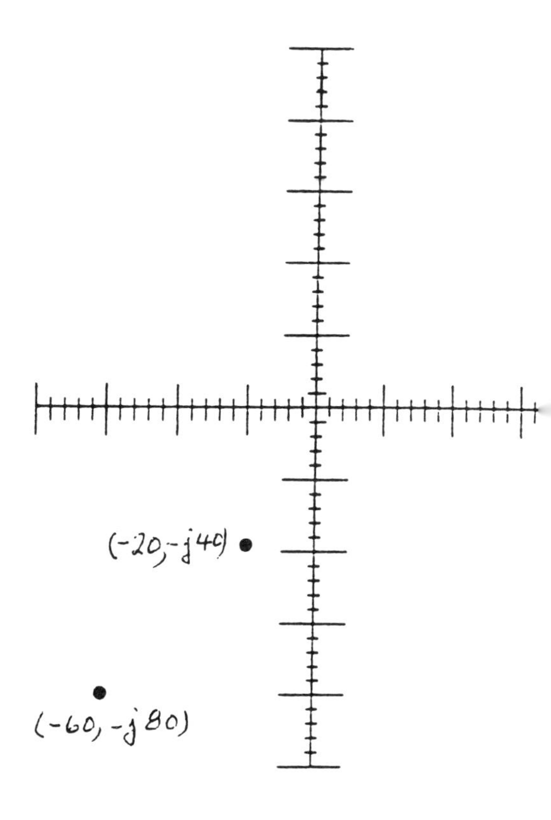

***Problems 16-3***

1. $15 + j27$
3. $25 + j25$
5. $27 - j8$
7. $30 + j10$
9. $20 + j2$
11. $20 - j8$
13. $10 - j6$
15. $17 - j10$
17. $-10 - j12$
19. $-21 - j20$
21. $5 + j3$
23. $-1 - j5$
25. $1 + j20$
27. $20 + j50$
29. $12 - j10$
31. $4 - j24$
33. $-2 + j0$
35. $3 - j2$
37. $2 + j2$
39. $-9 + j0$

***Problems 16-4***

1. $(5-j2)(3-j2) = 15 + j6 - j10 - j^2 4 = 19 - j4$
3. $(2 + j3)(2 - j3) = 4 + j6 - j6 - j^2 9 = 13$
5. $(4 - j3)(4 - j3) = 16 - j12 - j12 + j^2 9 = 7 - j24$
7. $(1 - j5)(3 - j3) = 3 - j15 - j3 + j^2 15 = -12 - j18$
9. $(4 + j5)(6 + j2) = 24 + j30 + j8 + j^2 10 = 14 + j38$
11. $(5 - j3)(6 + j1) = 30 - j18 + j5 - j^2 3 = 33 - j13$
13. $\dfrac{5 + j3}{1 + j4} \times \dfrac{1 - j4}{1 - j4} = \dfrac{5 + j3 - j20 - j^2 12}{1 + j4 - j4 - j^2 16} = \dfrac{17 - j17}{17} = 1 - j1$

15. $\dfrac{4 + j4}{3 - j} \times \dfrac{3 + j}{3 + j} = \dfrac{12 + j4 + j12 + j^2 4}{9 - j3 + j3 - j^2} = \dfrac{8 + j16}{10} = 0.8 + j1.6$

17. $\dfrac{1 + j2}{5 + j2} \times \dfrac{5 - j2}{5 - j2} = \dfrac{5 + j10 - j2 - j^2 4}{25 + j10 - j10 - j^2 4} = \dfrac{9 + j8}{29} = 0.31 + j0.276$

19. $\dfrac{2 - j5}{-3 + j4} \times \dfrac{3 + j4}{3 + j4} = \dfrac{6 - j15 + j8 - j^2 20}{-9 + j12 - j12 + j^2 16} = \dfrac{26 - j7}{-25} = -1.04 + j0.280$

21. $\dfrac{3 + j4}{4 + j5} \times \dfrac{4 - j5}{4 - j5} = \dfrac{12 + j16 - j15 - j^2 20}{16 - j^2 25} = \dfrac{32 + j}{41} = 0.780 + J0.0244$

# CHAPTER 17

***Problems 17-1***

1. a) $\phi = 55°$
   b) side R is the greater side
   c) side R lies opposite $\underline{/\phi}$
   d) side R lies adjacent $\underline{/\theta}$

3. a) $Z = [R^2 + X^2]^{½} = 4.83$ kΩ
   b) $R = [Z^2 - X^2]^{½} = 487$ Ω
   c) $X = [Z^2 - R^2]^{½} = 98.9$ kΩ

5. a) $Y = [G^2 + B^2]^{½} = 2.99$ ms
   b) $G = [Y^2 - B^2]^{½} = 800$ μS
   c) $B = [Y^2 - G^2]^{½} = 425$ μS

***Problems 17-2***

1. $c = [a^2 + b^2]^{½} = 26.1$ Angle B is greater
3. $b = [c^2 - b^2]^{½} = 245$
5. $a = [c^2 - b^2]^{½} = 68.1$

*For the following problems:* $Z = [R^2 + X^2]^{½}$

7. Z = 976 Ω. Angle θ is greater.
9. Z = 731 kΩ. Angle θ is greater.
11. Z = 84.5 kΩ. Angle φ is greater.

*For the following problems:* $X = [Z^2 - R_2]^{½}$

13. X = 6.8 kΩ. Angle θ is greater.
15. X = 150 kΩ. Angle θ is greater.

*For the following problems:* $R = [Z^2 - X^2]^{½}$

17. R = 27 kΩ. Angle θ is greater.
19. R = 90.5 Ω. Angle φ is greater.

*For the following problems:* $Y = [G^2 + B^2]^{½}$

21. Y = 832 μS. Angle θ is greater.
23. Y = 4.37 mS. Angle φ is greater.

*For the following problems:* $B = [Y^2 - G^2]^{½}$

25. B = 21.3 μS. Angle φ is greater.
27. B = 4.54 mS. Angle φ is greater.

*For the following problems:* $G = [Y^2 - B^2]^{½}$

29. G = 556 μS. Angle φ is greater.
31. G = 13.3 mS. Angle θ is greater.

***Problems 17-3***

1. a) $\sin\theta = \frac{X}{Z}$, $\cos\theta = \frac{R}{Z}$, $\tan\theta$ is $\frac{X}{R}$

   b) $\sin\phi = \frac{R}{Z}$, $\cos\phi = \frac{X}{Z}$, $\tan\phi$ is $\frac{R}{X}$

***Problems 17-4***

1. sin 27.3° = 0.4586
3. sin 63.6° - 0.8957
5. cos 78.3° = 0.2028
7. cos 23.7° = 0.9157
9. tan 67.3° = 2.391
11. tan 40.3° = 0.8481

*Problems 17-5*

1. $\sin^{-1} 0.39715 = 23.4°$

3. $\sin^{-1} 0.87036 = 60.5°$

5. $\cos^{-1} 0.55630 = 56.2°$

7. $\cos^{-1} 0.78043 = 38.7°$

9. $\tan^{-1} 1.47146 = 55.8°$

*Problems 17-6*

1. $\theta = \tan^{-1} \frac{a}{b} = \tan^{-1} \frac{25}{40} = 32°.$ $\phi = 90° - \theta = 58°$

   $c = \frac{a}{\sin \theta} = \frac{25}{\sin 32°} = 47.2$

3. $\theta = 90° - \phi = 55°.$ $b = \frac{a}{\tan \theta} = \frac{200}{\tan 55°} = 140.$ $c = \frac{a}{\sin \theta} = \frac{200}{\sin 55°} = 244$

5. $a = c \sin \theta = 17.3 \times \sin 40° = 11.1,$ $b = c \cos \theta = 17.3 \times \cos 40° = 13.3$

7. $R = \frac{X}{\tan \theta} = \frac{17.6\ k\Omega}{\tan 22.3°} = 42.9\ k\Omega,$ $Z = \frac{X}{\sin \theta} = \frac{17.6\ k\Omega}{\sin 22.3°} = 46.4\ k\Omega$

9. $R = \frac{X}{\tan \theta} = \frac{650\ \Omega}{\tan 22.3°} = 1.58\ k\Omega,$ $Z = \frac{X}{\sin \theta} = \frac{650\ \Omega}{\sin 22.3°} = 1.71\ k\Omega$

11. $\theta = \tan^{-1} \frac{X}{R} = \tan^{-1} \frac{10\ k\Omega}{5.6\ k\Omega} = 60.8°,$ $Z = \frac{X}{\sin \theta} = \frac{10\ k\Omega}{\sin 60.8°} = 11.5\ k\Omega$

13. $\theta = \tan^{-1} \frac{X}{R} = \tan^{-1} \frac{19.7\ k\Omega}{27\ k\Omega} = 36.1°,$ $Z = \frac{X}{\sin \theta} = \frac{19.7\ k\Omega}{\sin 36.1°} = 33.4\ k\Omega$

15. $X = R \tan \theta = 20\ k\Omega \times \tan 30° = 11.5\ k\Omega,$ $Z = \frac{R}{\cos \theta} = \frac{20\ k\Omega}{\cos 30°} = 23.1\ k\Omega$

17. $X = R \tan \theta = 270\ k\Omega \times \tan 37.5° = 207\ k\Omega,$ $Z = \frac{R}{\cos \theta} = \frac{270\ k\Omega}{\cos 37.5°} = 340\ k\Omega$

19. $R = Z \cos \theta = 40\ k\Omega \times \cos 67.5° = 15.3\ k\Omega,$

    $X = Z \sin \theta = 40\ k\Omega \times \sin 67.5° = 37\ k\Omega$

21. $R = Z \cos \theta = 785\ \Omega \times \cos 40.3° = 599\ \Omega,$

    $X = Z \sin \theta = 785\ \Omega \times \sin 40.3° = 508\ \Omega$

23. $\theta = \cos^{-1} \frac{R}{Z} = \cos^{-1} \frac{12\ k\Omega}{22\ k\Omega} = 56.9°,$

    $X = Z \sin \theta = 22\ k\Omega \times \sin 56.9° = 18.4\ k\Omega$

25. $\theta = \cos^{-1} \frac{R}{Z} = \cos^{-1} \frac{1.8\ k\Omega}{2.55\ k\Omega} = 45.1°,$

    $X = Z \sin \theta = 2.55\ k\Omega \times \sin 45.1° = 1.81\ k\Omega$

27. $\theta = \sin^{-1} \frac{X}{Z} = \sin^{-1} \frac{17.5\ k\Omega}{38\ k\Omega} = 27.4°,$

    $R = Z \cos \theta = 38\ k\Omega \times \cos 27.4° = 33.7\ k\Omega$

29. $\theta = \sin^{-1} \frac{X}{Z} = \sin^{-1} \frac{15.3\ k\Omega}{26.8\ k\Omega} = 34.8°,$

    $R = Z \cos \theta = 26.8\ k\Omega \times \cos 34.8° = 22.0\ k\Omega$

31. $B = G \tan\theta = 60\ \mu S \times \tan 25° = 28\ \mu S$, $Y = \dfrac{G}{\cos\theta} = \dfrac{60\ \mu S}{\cos 25°} = 66.2\ \mu S$

33. $B = G \tan\theta = 135\ \mu S \times \tan 65° = 290\ \mu S$, $Y = \dfrac{G}{\cos\theta} = \dfrac{135\ \mu S}{\cos 65°} = 319\ \mu S$

35. $\theta = \tan^{-1}\dfrac{B}{G} = \tan^{-1}\dfrac{8.73\ mS}{12.5\ mS} = 34.9°$, $Y = \dfrac{B}{\sin\theta} = \dfrac{8.73\ mS}{\sin 34.9°} = 15.3\ mS$

37. $\theta = \tan^{-1}\dfrac{B}{G} = \tan^{-1}\dfrac{17.3\ \mu S}{12.7\ \mu S} = 53.7°$, $Y = \dfrac{B}{\sin\theta} = \dfrac{17.3\ \mu S}{\sin 53.7°} = 21.5\ \mu S$

39. $G = \dfrac{B}{\tan\theta} = \dfrac{35\ mS}{\tan 10.7°} = 185\ mS$, $Y = \dfrac{B}{\sin\theta} = \dfrac{35\ mS}{\sin 10.7°} = 189\ mS$

41. $G = \dfrac{B}{\tan\theta} = \dfrac{3.32\ mS}{\tan 27.4°} = 6.40\ mS$, $Y = \dfrac{B}{\sin\theta} = \dfrac{3.32\ mS}{\sin 27.5°} = 7.20\ mS$

43. $\cos^{-1}\dfrac{G}{Y} = \cos^{-1}\dfrac{6.35\ mS}{9.35\ mS} = 47.2°$,

$B = Y \sin\theta = 9.35\ mS \times \sin 47.2° = 6.86\ mS$

45. $\cos^{-1}\dfrac{G}{Y} = \cos^{-1}\dfrac{2.33\ mS}{3.30\ mS} = 45.1°$,

$B = Y \sin\theta = 3.3\ mS \times \sin 45.1° = 2.34\ mS$

47. $G = Y \cos\theta = 70\ \mu S \times \cos 32° = 59.4\ \mu S$,

$B = Y \sin\theta = 70\ \mu S \times \sin 32° = 37.1\ \mu S$

49. $G = Y \cos\theta = 6.57\ mS \times \cos 55° = 3.77\ mS$,

$B = Y \sin\theta = 6.57\ mS \times \sin 55° = 5.38\ mS$

51. $\theta = \sin^{-1}\dfrac{B}{Y} = \sin^{-1}\dfrac{7.5\ mS}{11.5\ mS} = 40.7°$

$G = Y \cos\theta = 11.5\ mS \times \cos 40.7° = 8.72\ mS$

53. $\theta = \sin^{-1}\dfrac{B}{Y} = \sin^{-1}\dfrac{170\ \mu S}{320\ \mu S} = 32.1°$

$G = Y \cos\theta = 320\ \mu S \times \cos 32.1° = 271\ \mu S$

# CHAPTER 18

***Problems 18-1***

1. $B = 180° - (A+C) = 180° - 120° = 60°$

$$a = \frac{c \sin A}{\sin C} = \frac{35 \times \sin 50°}{\sin 70°} = 28.5$$

$$b = \frac{c \sin B}{\sin C} = \frac{35 \times \sin 60°}{\sin 70°} = 32.3$$

3. $\sin B = \dfrac{b \sin C}{c} = \dfrac{60 \times \sin 35°}{100} = 0.344, \ B = \sin^{-1} 0.344 = 20.1°$

$A = 180° - (B+C) = 180° - 55.1° = 125°$

$$a = \frac{c \sin A}{\sin C} = \frac{100 \times \sin 125°}{\sin 35°} = 143$$

5. $\sin B = \dfrac{b \sin A}{a} = \dfrac{22 \times \sin 28°}{15} = 0.689, \ B = \sin^{-1} 0.689 = 43.5°$

$C = 180° - (A+B) = 180° - 71.5° = 109°$

$$c = \frac{a \sin C}{\sin A} = \frac{15 \times \sin 35°}{\sin 28°} = 30.3$$

*Second solution:* $B = 180° - 43.5° = 137°$

$C = 180° - (A+B) = 180° - 165° = 15.5°$

$$c = \frac{a \sin C}{\sin A} = \frac{15 \times \sin 15.5°}{\sin 28°} = 8.55$$

7. $C = 180° - (A+B) = 180° - 100° = 80°$

$$b = \frac{a \sin B}{\sin A} = \frac{10 \times \sin 70°}{\sin 30°} = 18.8$$

$$c = \frac{a \sin C}{\sin A} = \frac{10 \times \sin 80°}{\sin 30°} = 19.7$$

9. $\sin B = \dfrac{b \sin C}{c} = \dfrac{160 \times \sin 22°}{100} = 0.599, \ B = \sin^{-1} 0.599 = 36.8°$

$A = 180° - (B+C) = 180° - 58.8° = 121°$

$$a = \frac{c \sin A}{\sin C} = \frac{100 \times \sin 121°}{\sin 22°} = 228$$

*Second solution:* $B = 180° - 36.8° = 143°$

$A = 180° - (B+C) = 180° - 165° = 14.8°$

$$a = \frac{c \sin A}{\sin C} = \frac{100 \times \sin 14.8°}{\sin 22°} = 68.3$$

11. $\sin B = \dfrac{b \sin A}{a} = \dfrac{20 \times \sin 25°}{10} = 0.845, \ B = \sin^{-1} 0.845 = 57.7°$

$C = 180° - (A+B) = 180° - 82.7° = 97.3°$

$$c = \frac{a \sin C}{\sin A} = \frac{10 \times \sin 97.3°}{\sin 25°} = 23.5$$

*Second solution:* $B = 180° - 57.7° = 122°$

$C = 180° - (A+B) = 180° - 147° = 32.7°$

$$c = \frac{a \sin C}{\sin A} = \frac{10 \times \sin 32.7°}{\sin 25°} = 12.8$$

13. $A = 180° - (B+C) = 180° - 155° = 25°$

$$a = \frac{b \sin A}{\sin B} = \frac{55 \times \sin 25°}{\sin 50°} = 30.3$$

$$c = \frac{b \sin C}{\sin B} = \frac{55 \times \sin 105°}{\sin 50°} = 69.4$$

15. $\sin B = \frac{b \sin A}{a} = \frac{65 \times \sin 95°}{120} = 0.540$, $B = \sin^{-1} 0.540 = 32.7°$

$C = 180° - (A+B) = 180° - 128° = 52.3°$

$$c = \frac{a \sin C}{\sin A} = \frac{120 \times \sin 52.3°}{\sin 95°} = 95.4$$

17. $\sin B = \frac{b \sin C}{c} = \frac{6 \times \sin 34°}{4.7} = 0.714$, $B = \sin^{-1} 0.714 = 45.6°$

$A = 180° - (B+C) = 180° - 79.6° = 100°$

$$a = \frac{c \sin A}{\sin C} = \frac{4.7 \times \sin 100°}{\sin 34°} = 8.27$$

*Second solution:* $B = 180° - 45.6° = 135°$

$A = 180° - (B+C) = 180° - 168° = 11.5°$

$$a = \frac{c \sin A}{\sin c} = \frac{4.7 \times \sin 11.6°}{\sin 34°} = 1.68$$

19. $B = 180° - (A+C) = 180° - 170° = 10°$

$$a = \frac{c \sin A}{\sin C} = \frac{14 \times \sin 125°}{\sin 45°} = 16.2$$

$$b = \frac{c \sin B}{\sin C} = \frac{14 \times \sin 10°}{\sin 45°} = 3.44$$

21. $B = 180° - (A+C) = 180° - 140° = 40°$

$$a = \frac{c \sin A}{\sin C} = \frac{350 \times \sin 30°}{\sin 110°} = 186$$

$$b = \frac{c \sin B}{\sin C} = \frac{350 \times \sin 40°}{\sin 110°} = 239$$

23. $\sin C = \frac{c \sin B}{b} = \frac{65 \times \sin 28°}{40} = 0.763$, $C = \sin^{-1} 0.763 = 49.7°$

$A = 180° - (B+C) = 180° - 77.7° = 102°$

$$a = \frac{b \sin A}{\sin B} = \frac{40 \times \sin 102°}{\sin 28°} = 83.3$$

*Second solution:* $C = 180° - 49.7° = 130°$

$A = 180° - (B+C) = 180° - 158° = 21.7°$

$$a = \frac{b \sin A}{\sin B} = \frac{40 \times \sin 21.7°}{\sin 28°} = 31.5$$

25. $\sin C = \dfrac{c \sin B}{b} = \dfrac{275 \times \sin 100°}{350} = 0.774$, $C = \sin^{-1} 0.774 = 50.7°$

$A = 180° - (B+C) = 180° - 151° = 29.3°$

$$a = \frac{b \sin A}{\sin B} = \frac{350 \times \sin 29.3°}{\sin 100°} = 174$$

27. $A = 180° - (B+C) = 180° - 117° = 63°$

$$a = \frac{b \sin A}{\sin B} = \frac{17 \times \sin 63°}{\sin 95°} = 15.2$$

$$c = \frac{b \sin C}{\sin B} = \frac{17 \times \sin 22°}{\sin 95°} = 6.39$$

29. $A = 180° - (B+C) = 180° - 40° = 140°$

$$a = \frac{b \sin A}{\sin B} = \frac{95 \times \sin 140°}{\sin 20°} = 179$$

$$c = \frac{b \sin C}{\sin B} = \frac{95 \times \sin 20°}{\sin 20°} = 95$$

31. $\sin C = \dfrac{c \sin B}{b} = \dfrac{200 \times \sin 70°}{250} = 0.752$, $C = \sin^{-1} 0.752 = 48.7°$

$A = 180° - (B+C) = 180° - 119° = 61.3°$

$$a = \frac{b \sin A}{\sin B} = \frac{250 \times \sin 61.3°}{\sin 70°} = 233$$

33. $\sin C = \dfrac{c \sin B}{b} = \dfrac{100 \times \sin 120°}{150} = 0.577$, $C = \sin^{-1} 0.577 = 35.3°$

$A = 180° - (B+C) = 180° - 155° = 24.7°$

$$a = \frac{b \sin A}{\sin B} = \frac{150 \times \sin 24.7°}{\sin 120°} = 72.5$$

35. $\sin C = \dfrac{c \sin B}{b} = \dfrac{33 \times \sin 33°}{20} = 0.899$, $C = \sin^{-1} 0.899 = 64.0°$

$A = 180° - (B+C) = 180° - 97° = 83°$

$$a = \frac{b \sin A}{\sin B} = \frac{20 \times \sin 83°}{\sin 33°} = 36.4$$

*Second solution:* $C = 180° - 64° = 116°$

$A = 180° - (B+C) = 180° - 149° = 31°$

$$a = \frac{b \sin A}{\sin B} = \frac{20 \times \sin 31°}{\sin 33°} = 18.9$$

37. $B = 180° - (A+C) = 180° - 142° = 38°$

$$a = \frac{c \sin A}{\sin C} = \frac{110 \times \sin 47°}{\sin 95°} = 80.8$$

$$b = \frac{c \sin B}{\sin C} = \frac{110 \times \sin 38°}{\sin 95°} = 68.0$$

39. $B = 180° - (A+C) = 180° - 120° = 60°$

$$a = \frac{c \sin A}{\sin C} = \frac{4.3 \times \sin 75°}{\sin 45°} = 5.87$$

$$b = \frac{c \sin B}{\sin C} = \frac{4.3 \times \sin 60°}{\sin 45°} = 5.27$$

***Problems 18-2***

1. $$\cos A = \frac{a^2 - b^2 - c^2}{-2bc} = \frac{b^2 + c^2 - a^2}{2bc} = \frac{130^2 + 90^2 - 70^2}{2 \times 130 \times 90} = 0.859$$

$A = \cos^{-1} 0.859 = 30.8°$

$$\cos B = \frac{a^2 + c^2 - b^2}{2ac} = \frac{70^2 + 90^2 - 130^2}{2 \times 70 \times 90} = -0.310, \ \cos^{-1} -0.310$$

$A = 108°$

$C = 180° - (A+B) = 180° - 139° = 41.2°$

3. $$\cos A = \frac{b^2 + c^2 - a^2}{2bc} = \frac{100^2 + 200^2 - 250^2}{2 \times 100 \times 200} = -0.313, \ \cos^{-1} -0.313 = 108°$$

$$\cos B = \frac{a^2 + c^2 - b^2}{2ac} = \frac{250^2 + 200^2 - 100^2}{2 \times 250 \times 200} = 0.925, \ \cos^{-1} 0.925 = 22.3°$$

$C = 180° - (A+B) = 180° - 130° = 49.7°$

5. $a = [b^2 + c^2 - 2bc \cos A]^{½} = [60^2 + 80^2 - 2 \times 60 \times 80 \times \cos 95°]^{½} = 104$

$$\sin B = \frac{b \sin A}{a} = \frac{60 \times \sin 95°}{104} = 0.575, \ \sin^{-1} 0.575 = 35.1°$$

$C = 180° - (A+B) = 180° - 130° = 49.9°$

7. $a = [b^2 + c^2 - 2bc \cos A]^{½} = [12^2 + 36^2 - 2 \times 12 \times 36 \times \cos 80°]^{½} = 35.9$

$$\sin B = \frac{b \sin A}{a} = \frac{12 \times \sin 80°}{35.9} = 0.329, \ \sin^{-1} 0.329 = 19.2°$$

$C = 180° - (A+B) = 180° - 99.2° = 80.8°$

9. $b = [a^2 + c^2 - 2ac \cos B]^{½} = [500^2 + 400^2 - 2 \times 500 \times 400 \times \cos 100°]^{½} = 69$

$$\sin A = \frac{a \sin B}{b} = \frac{500 \times \sin 100°}{692} = 0.712, \ \sin^{-1} 0.712 = 45.3°$$

$C = 180° - (A+B) = 180° - 145° = 34.7°$

11. $\cos A = \dfrac{b^2 + c^2 - a^2}{2bc} = \dfrac{300^2 + 150^2 - 400^2}{2 \times 300 \times 150} = -0.528, \ \cos^{-1} -0.528 = 122°$

$\cos B = \dfrac{a^2 + c^2 - b^2}{2ac} = \dfrac{400^2 + 150^2 - 300^2}{2 \times 400 \times 150} = 0.771, \ \cos^{-1} 0.771 = 39.6°$

$C = 180° - (A+B) = 180° - 162° = 18.6°$

13. $\cos A = \dfrac{b^2 + c^2 - a^2}{2bc} = \dfrac{40^2 + 25^2 - 35^2}{2 \times 40 \times 25} = 0.500, \ \cos^{-1} -0.500 = 60°$

$\cos B = \dfrac{a^2 + c^2 - b^2}{2ac} = \dfrac{35^2 + 25^2 - 40^2}{2 \times 35 \times 25} = 0.143, \ \cos^{-1} 0.143 = 81.8°$

$C = 180° - (A+B) = 180° - 142° = 38.2°$

15. $b = [a^2 + c^2 - 2ac \cos B]^{½} = [8.4^2 + 10^2 - 2 \times 8.4 \times 10 \times \cos 30°]^{½} = 5.01$

$\sin A = \dfrac{a \sin B}{b} = \dfrac{8.4 \times \sin 30°}{5.01} = 0.839, \ \sin^{-1} 0.839 = 57.0°$

$C = 180° - (A+B) = 180° - 87° = 93°$

17. $b = [a^2 + c^2 - 2ac \cos B]^{½} = [40^2 + 62^2 - 2 \times 40 \times 62 \times \cos 105°]^{½} = 82$

$\sin A = \dfrac{a \sin B}{b} = \dfrac{40 \times \sin 105°}{82} = 0.471, \ \sin^{-1} 0.471 = 28.1°$

$C = 180° - (A+B) = 180° - 133° = 46.9°$

19. $a = [b^2 + c^2 - 2bc \cos A]^{½} = [600^2 + 400^2 - 2 \times 600 \times 400 \times \cos 95°]^{½} = 750$

$\sin B = \dfrac{b \sin A}{a} = \dfrac{600 \times \sin 95°}{750} = 0.797, \ \sin^{-1} 0.797 = 52.9°$

$C = 180° - (A+B) = 180° - 148° = 32.1°$

21. $a = [b^2 + c^2 - 2bc \cos A]^{½} = [10^2 + 15^2 - 2 \times 10 \times 15 \times \cos 120°]^{½} = 21.8$

$\sin B = \dfrac{b \sin A}{a} = \dfrac{10 \times \sin 120°}{21.8} = 0.397, \ \sin^{-1} 0.397 = 23.4°$

$C = 180° - (A+B) = 180° - 143° = 36.6°$

23. $\cos A = \dfrac{b^2 + c^2 - a^2}{2bc} = \dfrac{45^2 + 60^2 - 45^2}{2 \times 45 \times 60} = 0.667, \ \cos^{-1} 0.667 = 48.2°$

$\cos B = \dfrac{a^2 + c^2 - b^2}{2ac} = \dfrac{45^2 - 60^2 - 45^2}{2 \times 45 \times 60} = 0.667, \ \cos^{-1} 0.667 = 48.2°$

$C = 180° - (A+B) = 180° - 96.4° = 83.6°$

25. $\cos A = \dfrac{b^2 + c^2 - a^2}{2bc} = \dfrac{140^2 + 190^2 - 170^2}{2 \times 140 \times 190} = 0.504, \; \cos^{-1} 0.504 = 59.8°$

$\cos B = \dfrac{a^2 + c^2 - b^2}{2ac} = \dfrac{170^2 + 190^2 - 140^2}{2 \times 170 \times 190} = 0.703, \; \cos^{-1} 0.703 = 45.4°$

$C = 180° - (A+B) = 180° - 105° = 74.9°$

27. $\cos A = \dfrac{b^2 + c^2 - a^2}{2bc} = \dfrac{525^2 + 300^2 - 425^2}{2 \times 525 \times 300} = 0.587, \; \cos^{-1} 0.587 = 54°$

$\cos B = \dfrac{a^2 + c^2 - b^2}{2ac} = \dfrac{425^2 + 300^2 - 525^2}{2 \times 425 \times 300} = -0.0196, \; \cos^{-1} -0.0196 = 91.1°$

$C = 180° - (A+B) = 180° - 145° = 34.8°$

29. $\cos A = \dfrac{b^2 + c^2 - a^2}{2bc} = \dfrac{8.5^2 + 8.5^2 - 5.35^2}{2 \times 8.5 \times 8.5} = 0.802, \; \cos^{-1} 0.802 = 36.7°$

$\cos B = \dfrac{a^2 + c^2 - b^2}{2ac} = \dfrac{5.35^2 + 8.5^2 - 8.5^2}{2 \times 5.35 \times 8.5} = 0.315, \; \cos^{-1} 0.315 = 71.7°$

$C = 180° - (A+B) = 180° - 108° = 71.7°$

# CHAPTER 19

***Problems 19-1***

1. $T = \frac{1}{f} = \frac{1}{600Hz} = 1.67\ ms$

3. $T = \frac{1}{f} = \frac{1}{1.2kHz} = 833\ \mu s$

5. $T = \frac{1}{f} = \frac{1}{1.76MHz} = 568\ ns$

7. $f = \frac{1}{T} = \frac{1}{400\ \mu s} = 2.50\ kHz$

9. $f = \frac{1}{T} = \frac{1}{1.33ms} = 752\ Hz$

11. $f = \frac{1}{T} = \frac{1}{25}\mu s = 40.0\ kHz$

13. 0.436 rad

15. 1.57 rad

17. 3.14 rad

19. 1.31 rad

21. 10.5 rad

23. 8.37°

25. 28.6°

27. 90°

29. 100°

31. 573°

***Problems 19-2***

1. a) $W = 2\pi f = 2\pi \times 60\ Hz = 377\ rad/s = 21{,}600°/s$
   b) $W = 2\pi \times 150\ Hz = 942\ rad/s = 54{,}000°/s$
   c) $W = 2\pi \times 200\ Hz = 1260\ rad/s = 72{,}000°/s$
   d) $W = 2\pi \times 300\ Hz = 1880\ rad/s = 108{,}000°/s$
   e) $W = 2\pi \times 600\ Hz = 3770\ rad/s = 216{,}000°/s$
   f) $W = 2\pi \times 800\ Hz = 5030\ rad/s = 288{,}000°/s$
   g) $W = 2\pi \times 1.25\ kHz = 7850\ rad/s = 450{,}000°/s$

3. a) $2\pi f = rad/s,\ \frac{rads}{2\pi} = f,\ \frac{12\ rad/s}{2\pi} = f = 1.91\ Hz$

   b) $f = \frac{50\ rad/s}{2\pi} = 7.96\ Hz$

   c) $f = \frac{400\ rad/s}{2\pi} = 63.7\ Hz$

   d) $f = \frac{900\ rad/s}{2\pi} = 143\ Hz$

   e) $f = \frac{2500\ rad/s}{2\pi} = 398\ Hz$

   f) $f = \frac{5000\ rad/s}{2\pi} = 796\ Hz$

   g) $f = \frac{7000\ rad/s}{2\pi} = 1.11\ kHz$

5. a) $f = \frac{W}{360°} = \frac{10{,}000°/s}{360°} = 27.8$ Hz

b) $f = \frac{15{,}000°/s}{360°} = 41.7$ Hz

c) $f = \frac{50{,}000°/s}{360°} = 139$ Hz

d) $f = \frac{120{,}000°/s}{360°} = 333$ Hz

e) $f = \frac{200{,}000°/s}{360°} = 556$ Hz

f) $f = \frac{7.35 \times 10^{5}°/s}{360°} = 2.04$ kHz

g) $f = \frac{1.65 \times 10^{6}°/s}{360°} = 4.58$ kHz

7. a) $\theta = 2\pi\ ft = 2\pi \times 500\text{ Hz} \times 100\mu s = 0.314$ rad
$\theta = 360°\ ft = 360° \times 500\text{ Hz} \times 100\mu s = 18°$

b) $\theta = 2\pi \times 500\text{ Hz} \times 250\ \mu s = 0.785$ rad
$\theta = 360° \times 500\text{ Hz} \times 250\ \mu s = 45°$

c) $\theta = 2\pi \times 500\text{ Hz} \times 1\text{ ms} = 3.14$ rad
$\theta = 360° \times 500\text{ Hz} \times 1\text{ ms} = 180°$

d) $\theta = 2\pi \times 500\text{ Hz} \times 2.8\text{ ms} = 8.8$ rad
$\theta = 360° \times 500\text{ Hz} \times 2.8\text{ ms} = 504°$

9. a) $\theta = 2\pi \times 2.55\text{ kHz} \times 100\ \mu s = 1.6$ rad
$\theta = 360° \times 2.55\text{ kHz} \times 100\ \mu s = 91.8°$

b) $\theta = 2\pi \times 2.55\text{ kHz} \times 250\ \mu s = 4.01$ rad
$\theta = 360° \times 2.55\text{ kHz} \times 250\ \mu s = 230°$

c) $\theta = 2\pi \times 2.55\text{ kHz} \times 1\text{ ms} = 16$ rad
$\theta = 360° \times 2.55\text{ kHz} \times 1\text{ ms} = 918°$

d) $\theta = 2\pi \times 2.55\text{ kHz} \times 2.8\text{ ms} = 44.9$ rad
$\theta = 360° \times 2.55\text{ kHz} \times 2.8\text{ ms} = 2570°$

11. a) $f = \frac{\theta}{2\pi t} = \frac{1.8\text{ rad}}{2\pi \times 20\ \mu s} = 14.3$ kHz

b) $f = \frac{3.25\text{ rad}}{2\pi \times 500\ \mu s} = 1.03$ kHz

c) $f = \frac{8.73\text{ rad}}{2\pi \times 150\ \mu s} = 9.26$ kHz

d) $f = \frac{17.6\text{ rad}}{2\pi \times 45\ \mu s} = 62.2$ kHz

13. a) $f = \frac{145°}{360° \times 2.3\text{ mS}} = 175$ Hz

b) $f = \frac{310°}{360° \times 500\ \mu s} = 1.72$ kHz

c) $f = \frac{1100°}{360° \times 50\ \mu s} = 61.1$ kHz

d) $f = \frac{15{,}000°}{360° \times 25\ \mu s} = 1.67$ MHz

15. a) $t = \dfrac{\theta}{2\pi f} = \dfrac{1\ \text{rad}}{2\pi \times 5\ \text{kHz}} = 31.8\ \mu s$

b) $t = \dfrac{0.125\ \text{rad}}{2\pi \times 5\ \text{kHz}} = 3.98\ \mu s$

c) $t = \dfrac{4.73\ \text{rad}}{2\pi \times 5\ \text{kHz}} = 151\ \mu s$

d) $t = \dfrac{\theta}{360° f} = \dfrac{20°}{360° \times 5\ \text{kHz}} = 11.1\ \mu s$

e) $t = \dfrac{90°}{360° \times 5\ \text{kHz}} = 50\ \mu s$

f) $t = \dfrac{300°}{360° \times 5\ \text{kHz}} = 167\ \mu s$

17. a) $t = \dfrac{1\ \text{rad}}{2\pi \times 15\ \text{kHz}} = 10.6\ \mu s$

b) $t = \dfrac{0.125\ \text{rad}}{2\pi \times 15\ \text{kHz}} = 1.33\ \mu s$

c) $t = \dfrac{4.73\ \text{rad}}{2\pi \times 15\ \text{kHz}} = 50.2\ \mu s$

d) $t = \dfrac{20°}{360° \times 15\ \text{kHz}} = 3.70\ \mu s$

e) $t = \dfrac{90°}{360° \times 15\ \text{kHz}} = 16.7\ \mu s$

f) $t = \dfrac{300°}{360° \times 15\ \text{kHz}} = 55.6\ \mu s$

## *Problems 19-3*

1. a) $e = E_{pk} \sin (2\pi ft)$

$= 25\ V_{pk} \sin (2\pi \times 2.5\ \text{kHz} \times 10\ \mu s) = 3.91\ V_t$

b) $e = 25\ V_{pk} \sin (2\pi \times 2.5\ \text{kHz} \times 25\ \mu s) = 9.57\ V_t$

c) $e = 25\ V_{pk} \sin (2\pi \times 2.5\ \text{kHz} \times 150\ \mu s) = 17.7\ V_t$

d) $e = 25\ V_{pk} \sin (2\pi \times 2.5\ \text{kHz} \times 225\ \mu s) = -9.57\ V_t$

e) $e = 25\ V_{pk} \sin (2\pi \times 2.5\ \text{kHz} \times 300\ \mu s) = -25\ V_t$

f) $e = 25\ V_{pk} \sin (2\pi \times 2.5\ \text{kHz} \times 375\ \mu s) = -9.57\ V_t$

3. a) $e = 12\ V_{pk} \sin (2\pi \times 12.7\ \text{kHz} \times 10\ \mu s) = 8.59\ V_t$

b) $e = 12\ V_{pk} \sin (2\pi \times 12.7\ \text{kHz} \times 25\ \mu s) = 10.9\ V_t$

c) $e = 12\ V_{pk} \sin (2\pi \times 12.7\ \text{kHz} \times 150\ \mu s) = -6.75\ V_t$

d) $e = 12\ V_{pk} \sin (2\pi \times 12.7\ \text{kHz} \times 225\ \mu s) = -9.37\ V_t$

e) $e = 12\ V_{pk} \sin (2\pi \times 12.7\ \text{kHz} \times 300\ \mu s) = -11.2\ V_t$

f) $e = 12\ V_{pk} \sin (2\pi \times 12.7\ \text{kHz} \times 375\ \mu s) = -12.0\ V_t$

5. a) $i = I_{pk} \sin \theta = I_{pk} \sin (2\pi ft)$

$= 700\ \mu A_{pk} \times \sin (2\pi \times 50\ \text{kHz} \times 5\ \mu s) = 700\ \mu A_t$

b) $i = 700\ \mu A_{pk} \times \sin (2\pi \times 50\ \text{kHz} \times 10\ \mu s) = 0$

c) $i = 700\ \mu A_{pk} \times \sin (2\pi \times 50\ \text{kHz} \times 15\ \mu s) = -700\ \mu A_t$

d) $i = 700\ \mu A_{pk} \times \sin (2\pi \times 50\ \text{kHz} \times 20\ \mu s) = 0$

e) $i = 700\ \mu A_{pk} \times \sin (2\pi \times 50\ \text{kHz} \times 25\ \mu s) = 700\ \mu A_t$

f) $i = 700\ \mu A_{pk} \times \sin (2\pi \times 50\ \text{kHz} \times 30\ \mu s) = 0$

7. a) $i = 3.45\ mA_{pk} \times \sin(2\pi \times 800\ Hz \times 5\ \mu s) = 86.7\ \mu A_t$

   b) $i = 3.45\ mA_{pk} \times \sin(2\pi \times 800\ Hz \times 10\ \mu s) = 173\ \mu A_t$

   c) $i = 3.45\ mA_{pk} \times \sin(2\pi \times 800\ Hz \times 15\ \mu s) = 260\ \mu At$

   d) $i = 3.45\ mA_{pk} \times \sin(2\pi \times 800\ Hz \times 20\ \mu s) = 346\ \mu A_t$

   e) $i = 3.45\ mA_{pk} \times \sin(2\pi \times 800\ Hz \times 25\ \mu s) = 432\ \mu A_t$

   f) $i = 3.45\ mA_{pk} \times \sin(2\pi \times 800\ Hz \times 30\ \mu s) = 518\ \mu A_t$

***Problems 19-4***

1. a) $65\ V \times 1.414 = 91.9\ V_{pk}$, $91.9\ V_{pk} \times 2 = 184\ V_{p-p}$

   b) $110\ V \times 1.414 = 156\ V_{pk}$, $156\ V_{pk} \times 2 = 311\ V_{p-p}$

3. a) $220\ V \times 1.414 = 311\ V_{pk}$, $311\ V_{pk} \times 2 = 622\ V_{p-p}$

   b) $440\ V \times 1.414 = 622\ V_{pk}$, $622\ V_{pk} \times 2 = 1.24\ kV_{p-p}$

5. a) $600\ \mu A \times 1.414 = 848\ \mu A_{pk}$, $848\ \mu A_{pk} \times 2 = 1.70\ mA_{p-p}$

   b) $1.2\ mA \times 1.414 = 1.70\ mA_{pk}$, $1.70\ mA_{pk} \times 2 = 3.39\ mA_{p-p}$

7. a) $80\ V_{pk} \times 2 = 160\ V_{p-p}$, $80\ V_{pk} \times 0.707 = 56.6\ V$

   b) $3\ A = 1.414 \times 3A = 4.24\ A_{pk}$, $4.24\ A_{pk} \times 2 = 8.48\ A_{p-p}$

9. a) $350\ \mu V_{pk} \times 2 = 700\ \mu V_{p-p}$, $350\ \mu V_{pk} \times 0.707 = 247\ \mu V$

   b) $400\ mV_{pk} \times 2 = 800\ mV_{p-p}$, $400\ mV_{pk} \times 0.707 = 283\ mV$

11. a) $80\ V_{pk} \times 2 = 160\ V_{p-p}$, $80\ V_{pk} \times 0.707 = 56.6\ V$

    b) $200\ V_{pk} \times 2 = 400\ V_{p-p}$, $200\ V_{pk} \times 0.707 = 141\ V$

13. a) $200\ \mu A_{pk} \times 2 = 400\ \mu A_{p-p}$, $200\ \mu A_{pk} \times 0.707 = 141\ \mu A$

    b) $9.45\ mA_{pk} \times 2 = 18.9\ mA_{p-p}$, $9.45\ mA_{pk} \times 0.707 = 6.68\ mA$

15. a) $500\ mA_{pk} \times 2 = 1\ A_{p-p}$, $500\ mA_{pk} \times 0.707 = 354\ mA$

    b) $1.75\ A_{pk} \times 2 = 3.50\ A_{p-p}$, $1.75\ A_{pk} \times 0.707 = 1.24\ A$

17. a) $V_{pk} = \frac{320\ \mu V_{p-p}}{2} = 160\ V_{pk}$, $320\ V_{p-p} \times 0.3535 = 113\ \mu V$

    b) $V_{pk} = \frac{70\ mV_{p-p}}{2} = 35\ V_{pk}$, $70\ V_{p-p} \times 0.3535 = 24.7\ mV$

19. a) $V_{pk} = \frac{17\ V_{p-p}}{2} = 8.5\ V_{pk}$, $17\ V_{p-p} \times 0.3535 = 6.01\ V$

    b) $V_{pk} = \frac{283\ V_{p-p}}{2} = 142\ V_{pk}$, $283\ V_{p-p} \times 0.3535 = 100\ V$

21. a) $\mu A_{pk} = \frac{50\ \mu A_{p-p}}{2} = 25\ \mu A_{pk}$, $50\ \mu A_{p-p} \times 0.3535 = 17.7\ \mu A$

    b) $\mu A_{pk} = \frac{700\ \mu A_{p-p}}{2} = 350\ \mu A_{pk}$, $700\ \mu A_{pk} \times 0.3535 = 247\ \mu A$

23. a) $A_{p\text{-}p} = \dfrac{9.38\ mA_{p\text{-}p}}{2} = 4.69\ mA_{pk}$, $9.38\ mA_{p\text{-}p} \times 0.3535 = 3.32\ mA$

b) $A_{p\text{-}p} = \dfrac{650\ mA_{p\text{-}p}}{2} = 325\ mA_{pk}$, $650\ mA_{p\text{-}p} \times 0.3535 = 230\ mA$

***Problems 19-5***

1. a) $\theta = wt = 360° \times 20\ kHz \times 5\ \mu s = 36°$
   $i = I_{pk} \sin\theta = 4\ mA_{pk} \sin 36° = 2.35\ mA_t$
   $e = E_{pk} \sin(\theta + 25°) = 10\ V_{pk} \sin(36° + 25°) = 8.75\ V_t$
   b) $\theta = 86.4°$
   $i = 4\ mA_{pk} \sin 86.4° = 3.99\ mA_t$
   $e = 10\ V_{pk} \sin(86.4° + 25°) = 9.31\ V_t$
   c) $\theta = 180°$
   $i = 4\ mA_{pk} \sin 180° = 0$
   $e = 10\ V_{pk} \sin(180° + 25°) = -4.23\ V_t$
   d) $\theta = 288°$
   $i = 4\ mA_{pk} \sin 288° = -3.80\ mA_t$
   $e = 10\ v_{pk} \sin(288° + 25°) = -7.31\ V_t$

3. a) $\theta = wt = 360° \times 100\ Hz \times 1\ ms = 36°$
   $i = 25\ mA_{pk} \times \sin(36° + 35°) = 23.6\ mA_t$
   $e = 28.3\ V_{pk} \times \sin 36° = 16.6\ V_t$
   b) $\theta = 90°$
   $i = 25\ mA_{pk} \times \sin(90° + 35°) = 20.5\ mA_t$
   $e = 28.3\ V_{pk} \times \sin 90° = 28.3\ V_t$
   c) $\theta = 115°$
   $i = 25\ mA_{pk} \times \sin(115° + 35°) = 12.4\ mA_t$
   $e = 28.3\ V_{pk} \times \sin 115° = 25.6\ V_t$
   d) $\theta = 299°$
   $i = 25\ mA_{pk} \times \sin(299° + 35°) = -11.0\ mA_t$
   $e = 28.3\ V_{pk} \times \sin 299° = -24.8\ V_t$

5. a) $\theta = wt = 360° \times 3.75\ kHz \times 10\ \mu s = 13.5°$
   $i = 600\ \mu A_{pk} \sin 13.5° = 140\ \mu A_t$
   $e = 80\ V_{pk} \sin(13.5° - 55°) = -53\ V_t$
   b) $\theta = 54°$
   $i = 600\ \mu A_{pk} \sin 54° = 485\ \mu A_t$
   $e = 80\ V_{pk} \sin(54° - 55°) = -1.40\ V_t$
   c) $\theta = 162°$
   $i = 600\ \mu A_{pk} \sin 162° = 185\ \mu A_t$
   $e = 80\ V_{pk} \sin(162° - 55°) = 76.5\ V_t$

d) $\theta = 270°$

$i = 600\ \mu A_{pk} \sin 270° = -600\ \mu A_t$

$e = 80\ V_{pk} \sin (270° - 55°) = -45.9\ V_t$

7. a) $\theta = wt = 360° \times 2.35\ kHz \times 25\ \mu s = 21.2°$

$i = 25\ mA_{pk} \times \sin (21.2° + 40°) = 21.9\ mA_t$

$e = 65\ V_{pk} \times \sin 21.2° = 23.5\ V_t$

b) $\theta = 59.2°$

$i = 25\ mA_{pk} \times \sin (59.2° + 40°) = 24.7\ mA_t$

$e = 65\ V_{pk} \times \sin 59.2° = 55.8\ V_t$

c) $\theta = 169°$

$i = 25\ mA_{pk} \times \sin (169° + 40°) = -12.2\ mA_t$

$e = 65\ V_{pk} \times \sin 169° = 12.2\ V_t$

d) $\theta = 508°$

$i = 25\ mA_{pk} \times \sin (508° + 40°) = -3.31\ mA_t$

$e = 65\ V_{pk} \times \sin 508° = 34.8\ V_t$

# CHAPTER 20

***Problems 20-1***

1. $\theta = \tan^{-1}\dfrac{V_C}{V_R} = \tan^{-1}\dfrac{-20\text{ V}}{20\text{ V}} = -45°$

   $E = \dfrac{V_C}{\sin\theta} = \dfrac{-20\text{ V}}{\sin -45°} = 28.3\text{ V}$

3. $\theta = \tan^{-1}\dfrac{V_C}{V_R} = \tan^{-1}\dfrac{-8\text{ V}}{12\text{ V}} = -33.7°$

   $E = \dfrac{V_C}{\sin\theta} = \dfrac{-8\text{ V}}{\sin -33.7°} = 14.4\text{ V}$

5. $\theta = \cos^{-1}\dfrac{V_R}{E} = \cos^{-1}\dfrac{6.35\text{ V}}{10\text{ V}} = -50.6°$

   $V_C = E\sin\theta = 10\text{ V} \times \sin -50.6° = -7.73\text{ V}$

7. $\theta = \cos^{-1}\dfrac{V_R}{E} = \cos^{-1}\dfrac{21\text{ V}}{40\text{ V}} = -58.3°$

   $V_C = E\sin\theta = 40\text{ V} \times \sin -58.3° = -34\text{ V}$

9. $E = \dfrac{V_R}{\cos\theta} = \dfrac{15\text{ V}}{\cos -40°} = 19.6\text{ V}$

   $V_C = V_R\tan\theta = 15\text{ V} \times \tan -40° = -12.6\text{ V}$

11. $E = \dfrac{V_R}{\cos\theta} = \dfrac{5.5\text{ V}}{\cos -25°} = 6.07\text{ V}$

    $V_C = V_R\tan\theta = 5.5\text{ V} \times \tan -25° = -2.56\text{ V}$

13. $V_R = E\cos\theta = 18\text{ V} \times \cos -35.6° = 14.6\text{ V}$

    $V_C = E\sin\theta = 18\text{ V} \times \sin -35.6° = -10.5\text{ V}$

15. $V_R = E\cos\theta = 70\text{ V} \times \cos -27° = 62.4\text{ V}$

    $V_C = E\sin\theta = 70\text{ V} \times \sin -27° = -31.8\text{ V}$

17. $\theta = \sin^{-1}\dfrac{V_C}{E} = \sin^{-1}\dfrac{-20\text{ V}}{30\text{ V}} = -41.8°$

    $V_R = E\cos\theta = 30\text{ V} \times \cos -41.8° = 22.4\text{ V}$

19. $\theta = \sin^{-1}\dfrac{V_C}{E} = \sin^{-1}\dfrac{-40\text{ V}}{50\text{ V}} = -53.1°$

    $V_R = E\cos\theta = 50\text{ V} \times \cos -53.1° = 30\text{ V}$

21. $E = \dfrac{V_C}{\sin\theta} = \dfrac{5.5\ V}{\sin -32°} = 10.4\ V$

$V_R = \dfrac{V_C}{\tan\theta} = \dfrac{-5.5\ V}{\tan -32°} = 8.82\ V$

23. $E = \dfrac{V_C}{\sin\theta} = \dfrac{-30\ V}{\sin -70°} = 31.9\ V$

$V_R = \dfrac{V_C}{\tan\theta} = \dfrac{-30\ V}{\tan -70°} = 10.9\ V$

***Problems 20-2***

1. $X_C = \dfrac{1}{2\pi fC} = \dfrac{1}{2\pi \times 500\ Hz \times 50\ nF} = 6.37\ k\Omega$

$\theta = \tan^{-1} \dfrac{X_C}{R} = \tan^{-1} \dfrac{6.37\ k\Omega}{10\ k\Omega} = -32.5°$

$Z = \dfrac{R}{\cos\theta} = \dfrac{10\ k\Omega}{\cos -32.5°} = 11.9\ k\Omega$

3. $X_C = \dfrac{1}{2\pi fC} = \dfrac{1}{2\pi \times 200\ Hz \times 1\ \mu F} = 796\ \Omega$

$\theta = \tan^{-1} \dfrac{X_C}{R} = \tan^{-1} \dfrac{796\ k\Omega}{700\ \Omega} = -48.7°$

$Z = \dfrac{R}{\cos\theta} = \dfrac{700\ \Omega}{\cos -48.7°} = 1.06\ k\Omega$

5. $X_C = \dfrac{1}{2\pi fC} = \dfrac{1}{2\pi \times 2.5\ kHz \times 20\ nF} = 3.18\ k\Omega$

$\theta = \tan^{-1} \dfrac{X_C}{R} = \tan^{-1} \dfrac{3.18\ k\Omega}{2\ k\Omega} = -57.9°$

$Z = \dfrac{R}{\cos\theta} = \dfrac{2\ k\Omega}{\cos -57.9°} = 3.76\ k\Omega$

7. $X_C = \dfrac{1}{2\pi fC} = \dfrac{1}{2\pi \times 1.55\ kHz \times 5\ nF} = 20.5\ k\Omega$

$\theta = \tan^{-1} \dfrac{X_C}{R} = \tan^{-1} \dfrac{20.5\ k\Omega}{27\ k\Omega} = -37.3°$

$Z = \dfrac{R}{\cos\theta} = \dfrac{27\ k\Omega}{\cos -32.5°} = 33.9\ k\Omega$

9. $X_C = \dfrac{1}{2\pi fC} = \dfrac{1}{2\pi \times 2.3\ kHz \times 0.1\ \mu F} = 692\ \Omega$

$\theta = \tan^{-1} \dfrac{X_C}{R} = \tan^{-1} \dfrac{692\ \Omega}{3.3\ k\Omega} = -11.8°$

$Z = \dfrac{R}{\cos\theta} = \dfrac{3.3\ k\Omega}{\cos -11.8°} = 3.37\ k\Omega$

11. $X_C = \frac{1}{2\pi fC} = \frac{1}{2\pi \times 3\ kHz \times 40\ nF} = 1.33\ k\Omega$

$\theta = \tan^{-1} \frac{X_C}{R} = \tan^{-1} \frac{1.33\ k\Omega}{2\ k\Omega} = -33.6°$

$Z = \frac{R}{\cos\theta} = \frac{2\ k\Omega}{\cos-33.6°} = 2.40\ k\Omega$

$V_R = E\cos\theta = 30\ V \times \cos-33.6° = 25\ V$

$V_C = E\sin\theta = 30\ V \times \sin-33.6° = -16.6\ V$

$I = \frac{E}{Z} = \frac{30\ V}{2.4\ k\Omega} = 12.5\ mA$

13. $X_C = \frac{1}{2\pi fC} = \frac{1}{2\pi \times 100\ kHz \times 100\ pF} = 15.9\ k\Omega$

$\theta = \tan^{-1} \frac{X_C}{R} = \tan^{-1} \frac{15.9\ k\Omega}{10\ k\Omega} = -57.9°$

$Z = \frac{R}{\cos\theta} = \frac{10\ k\Omega}{\cos-57.9°} = 18.8\ k\Omega$

$V_R = E\cos\theta = 10\ V \times \cos-57.9° = 5.32\ V$

$V_C = E\sin\theta = 10\ V \times \sin-57.9° = -8.47\ V$

$I = \frac{E}{Z} = \frac{10\ V}{18.8\ k\Omega} = 532\ \mu A$

15. $X_C = \frac{1}{2\pi fC} = \frac{1}{2\pi \times 1.5\ kHz \times 20\ nF} = 5.31\ k\Omega$

$\theta = \tan^{-1} \frac{X_C}{R} = \tan^{-1} \frac{5.31\ k\Omega}{3.3\ k\Omega} = -58.1°$

$Z = \frac{R}{\cos\theta} = \frac{3.3\ k\Omega}{\cos-58.1°} = 6.25\ k\Omega$

$V_R = E\cos\theta = 30\ V \times \cos-58.1° = 15.8\ V$

$V_C = E\sin\theta = 30\ V \times \sin-58.1° = -25.5\ V$

$I = \frac{E}{Z} = \frac{30\ V}{6.25\ k\Omega} = 4.8\ mA$

17. $X_C = \frac{1}{2\pi fC} = \frac{1}{2\pi \times 6.5\ kHz \times 500\ pF} = 49.0\ k\Omega$

$\theta = \tan^{-1} \frac{X_C}{R} = \tan^{-1} \frac{49\ k\Omega}{10\ k\Omega} = -78.5°$

$Z = \frac{R}{\cos\theta} = \frac{10\ k\Omega}{\cos-78.5°} = 50.0\ k\Omega$

$V_R = E\cos\theta = 20\ V \times \cos-78.5° = 4\ V$

$V_C = E\sin\theta = 20\ V \times \sin-78.5° = -19.6\ V$

$I = \frac{E}{Z} = \frac{20\ V}{50\ k\Omega} = 400\ \mu A$

19. $X_C = \frac{1}{2\pi fC} = \frac{1}{2\pi \text{ x } 2 \text{ kHz x } 200 \text{ nF}} = 398\ \Omega$

$\theta = \tan^{-1} \frac{X_C}{R} = \tan^{-1} \frac{398\ \Omega}{470\ \Omega} = -40.3°$

$Z = \frac{R}{\cos\theta} = \frac{470\ \Omega}{\cos -40.3°} = 616\ \Omega$

$V_R = E\cos\theta = 100 \text{ V x } \cos -40.3° = 76.3 \text{ V}$

$V_C = E\sin\theta = 100 \text{ V x } \sin -40.3° = -64.6 \text{ V}$

$I = \frac{E}{Z} = \frac{100 \text{ V}}{616\ \Omega} = 162 \text{ mA}$

21. $\theta = \cos^{-1} \frac{R}{Z} = \cos^{-1} \frac{8.2 \text{ k}\Omega}{12 \text{ k}\Omega} = -46.9°$

$X_C = Z\sin\theta = 12 \text{ k}\Omega \text{ x } \sin -46.9° = 8.76 \text{ k}\Omega$

$f = \frac{1}{2\pi X_C C} = \frac{1}{2\pi \text{ x } 8.76 \text{ k}\Omega \text{ x } 200 \text{ pF}} = 90.8 \text{ kHz}$

$V_R = E\cos\theta = 40 \text{ V x } \cos -46.9° = 27.3 \text{ V}$

$V_C = E\sin\theta = 40 \text{ V x } \sin -46.9° = -29.2 \text{ V}$

$I = \frac{E}{Z} = \frac{40 \text{ V}}{12 \text{ k}\Omega} = 3.33 \text{ mA}$

23. $\theta = \cos^{-1} \frac{R}{Z} = \cos^{-1} \frac{750\ \Omega}{1.2 \text{ k}\Omega} = -51.3°$

$X_C = Z\sin\theta = 1.2 \text{ k}\Omega \text{ x } \sin -51.3° = 937 \text{ k}\Omega$

$f = \frac{1}{2\pi X_C C} = \frac{1}{2\pi \text{ x } 937\ \Omega \text{ x } 10 \text{ nF}} = 17 \text{ kHz}$

$V_R = E\cos\theta = 15 \text{ V x } \cos -51.3° = 9.38 \text{ V}$

$V_C = E\sin\theta = 15 \text{ V x } \sin -51.3° = -11.7 \text{ V}$

$I = \frac{E}{Z} = \frac{15 \text{ V}}{1.2 \text{ k}\Omega} = 12.5 \text{ mA}$

25. $\theta = \cos^{-1} \frac{R}{Z} = \cos^{-1} \frac{2.7 \text{ k}\Omega}{3.5 \text{ k}\Omega} = -39.5°$

$X_C = Z\sin\theta = 3.5 \text{ k}\Omega \text{ x } \sin -39.5° = 2.23 \text{ k}\Omega$

$f = \frac{1}{2\pi X_C C} = \frac{1}{2\pi \text{ x } 2.23 \text{ k}\Omega \text{ x } 30 \text{ pF}} = 2.38 \text{ kHz}$

$V_R = E\cos\theta = 30 \text{ V x } \cos -39.5° = 23.1 \text{ V}$

$V_C = E\sin\theta = 30 \text{ V x } \sin -39.5° = -19.1 \text{ V}$

$I = \frac{E}{Z} = \frac{30 \text{ V}}{3.5 \text{ k}\Omega} = 8.57 \text{ mA}$

27. $\theta = \cos^{-1} \frac{R}{Z} = \cos^{-1} \frac{56 \text{ k}\Omega}{64.4 \text{ k}\Omega} = -29.6°$

$X_C = Z\sin\theta = 64.4 \text{ k}\Omega \text{ x } \sin -29.6° = 31.8 \text{ k}\Omega$

$f = \frac{1}{2\pi X_C C} = \frac{1}{2\pi \text{ x } 31.8 \text{ k}\Omega \text{ x } 25 \text{ nF}} = 200 \text{ Hz}$

$V_R = E\cos\theta = 7.5 \text{ V x } \cos -29.6° = 6.52 \text{ V}$

$V_C = E\sin\theta = 7.5\text{ V} \times \sin -29.6° = -3.70\text{ V}$

$I = \frac{E}{Z} = \frac{7.5\text{ V}}{64.6\text{ k}\Omega} = 116\ \mu\text{A}$

29. $\theta = \cos^{-1}\frac{R}{Z} = \cos^{-1}\frac{810\ \Omega}{1.14\text{ k}\Omega} = -44.7°$

$X_C = Z\sin\theta = 1.14\text{ k}\Omega \times \sin -44.7° = 802\ \Omega$

$f = \frac{1}{2\pi X_C C} = \frac{1}{2\pi \times 802\ \Omega \times 1\ \mu\text{F}} = 198\text{ Hz}$

$V_R = E\cos\theta = 12\text{ V} \times \cos -44.7° = 8.53\text{ V}$

$V_C = E\sin\theta = 12\text{ V} \times \sin -44.7° = -8.44\text{ V}$

$I = \frac{E}{Z} = \frac{12\text{ V}}{1.14\text{ k}\Omega} = 10.5\text{ mA}$

***Problems 20-3***

1. $\theta = \tan^{-1}\frac{V_L}{V_R} = \tan^{-1}\frac{10\text{ V}}{8\text{ V}} = 51.3°$

$E = \frac{V_L}{\sin\theta} = \frac{10\text{ V}}{\sin 45°} = 12.8\text{ V}$

3. $\theta = \tan^{-1}\frac{V_L}{V_R} = \tan^{-1}\frac{20\text{ V}}{14\text{ V}} = 55°$

$E = \frac{V_L}{\sin\theta} = \frac{20\text{ V}}{\sin 55°} = 24.4\text{ V}$

5. $E = \frac{V_R}{\cos\theta} = \frac{6.7\text{ V}}{\cos 27°} = 7.52\text{ V}$

$V_L = E\sin\theta = 7.52 \times \sin 27° = 3.41\text{ V}$

7. $E = \frac{V_R}{\cos\theta} = \frac{27.4\text{ V}}{\cos 67.4°} = 71.3\text{ V}$

$V_L = E\sin\theta = 71.3 \times \sin 67.4° = 65.8\text{ V}$

9. $\theta = \cos^{-1}\frac{V_R}{E} = \cos^{-1}\frac{8.35\text{ V}}{12\text{ V}} = 45.9°$

$V_L = E\sin\theta = 12\text{ V} \times \sin 45.9° = 8.62\text{ V}$

11. $\theta = \cos^{-1}\frac{V_R}{E} = \cos^{-1}\frac{43.7\text{ V}}{60\text{ V}} = 43.3°$

$V_L = E\sin\theta = 60\text{ V} \times \sin 43.3° = 41.1\text{ V}$

13. $\theta = \sin^{-1}\frac{V_L}{E} = \sin^{-1}\frac{15\text{ V}}{30\text{ V}} = 30°$

$V_R = E\cos\theta = 30\text{ V} \times \cos 30° = 26\text{ V}$

15. $\theta = \sin^{-1}\dfrac{V_L}{E} = \sin^{-1}\dfrac{10\ V}{17\ V} = 36°$

$V_R = E\cos\theta = 17\ V \times \cos 36° = 13.8\ V$

17. $E = \dfrac{V_L}{\sin\theta} = \dfrac{3\ V}{\sin 62°} = 3.4\ V$

$V_R = \dfrac{V_L}{\tan\theta} = \dfrac{3\ V}{\tan 62°} = 1.6\ V$

19. $E = \dfrac{V_L}{\sin\theta} = \dfrac{15.5\ V}{\sin 30°} = 31\ V$

$V_R = \dfrac{V_L}{\tan\theta} = \dfrac{15.5\ V}{\tan 30°} = 26.8\ V$

21. $V_R = E\cos\theta = 9\ V \times \cos 58° = 4.77\ V$

$V_L = E\sin\theta = 9\ V \times \sin 58° = 7.63\ V$

23. $V_R = E\cos\theta = 50\ V \times \cos 45° = 35.4\ V$

$V_L = E\sin\theta = 50\ V \times \sin 45° = 35.4\ V$

***Problems 20-4***

1. $X_L = 2\pi fL = 2\pi \times 1\ kHz \times 600\ mH = 3.77\ k\Omega$

$\theta = \tan^{-1}\dfrac{X_L}{R} = \tan^{-1}\dfrac{3.77\ k\Omega}{3.9\ k\Omega} = 44°$

$Z = \dfrac{X_L}{\sin\theta} = \dfrac{3.77\ k\Omega}{\sin 44°} = 5.43\ k\Omega$

3. $X_L = 2\pi fL = 2\pi \times 1.5\ kHz \times 40\ mH = 377\ \Omega$

$\theta = \tan^{-1}\dfrac{X_L}{R} = \tan^{-1}\dfrac{377\ \Omega}{500\ \Omega} = 37°$

$Z = \dfrac{X_L}{\sin\theta} = \dfrac{377\ \Omega}{\sin 37°} = 626\ \Omega$

5. $X_L = 2\pi fL = 2\pi \times 8.5\ kHz \times 250\ mH = 13.4\ k\Omega$

$\theta = \tan^{-1}\dfrac{X_L}{R} = \tan^{-1}\dfrac{13.4\ k\Omega}{27\ k\Omega} = 26.3°$

$Z = \dfrac{X_L}{\sin\theta} = \dfrac{13.4\ k\Omega}{\sin 26.3°} = 30\ k\Omega$

7. $X_L = 2\pi fL = 2\pi \times 20\ kHz \times 10\ mH = 1.26\ k\Omega$

$\theta = \tan^{-1}\dfrac{X_L}{R} = \tan^{-1}\dfrac{1.26\ k\Omega}{1.8\ k\Omega} = 34.9°$

$Z = \dfrac{X_L}{\sin\theta} = \dfrac{1.26\ k\Omega}{\sin 34.9°} = 2.2\ k\Omega$

9. $X_L = 2\pi fL = 2\pi \times 250\ Hz \times 500\ mH = 785\ \Omega$

$$\theta = \tan^{-1}\frac{X_L}{R} = \tan^{-1}\frac{785\ \Omega}{1\ k\Omega} = 38.1°$$

$$Z = \frac{X_L}{\sin\theta} = \frac{785\ \Omega}{\sin 38.1°} = 1.27\ k\Omega$$

$V_R = E\cos\theta = 15\ V \times \cos 38.1° = 11.8\ V$

$V_L = E\sin\theta = 15\ V \times \sin 38.1° = 9.27\ V$

$$I = \frac{E}{Z} = \frac{15\ V}{1.27\ k\Omega} = 11.8\ mA$$

11. $X_L = 2\pi fL = 2\pi \times 500\ Hz \times 150\ mH = 471\ \Omega$

$$\theta = \tan^{-1}\frac{X_L}{R} = \tan^{-1}\frac{471\ \Omega}{300\ \Omega} = 57.5°$$

$$Z = \frac{X_L}{\sin\theta} = \frac{471\ \Omega}{\sin 57.5°} = 559\ \Omega$$

$V_R = E\cos\theta = 50\ V \times \cos 57.5° = 26.9\ V$

$VL = E\sin\theta = 50\ V \times \sin 57.5° = 42.2\ V$

$$I = \frac{E}{Z} = \frac{50\ V}{559} = 89.4\ mA$$

13. $X_L = 2\pi fL = 2\pi \times 2.35\ kHz \times 1.35\ H = 19.9\ k\Omega$

$$\theta = \tan^{-1}\frac{X_L}{R} = \tan^{-1}\frac{19.9\ k\Omega}{27\ k\Omega} = 36.5°$$

$$Z = \frac{X_L}{\sin\theta} = \frac{19.9\ k\Omega}{\sin 36.5°} = 33.6\ k\Omega$$

$V_R = E\cos\theta = 20\ V \times \cos 36.4° = 16.1\ V$

$V_L = E\sin\theta = 20\ V \times \sin 36.4° = 11.9\ V$

$$I = \frac{E}{Z} = \frac{20\ V}{33.6\ k\Omega} = 595\ mA$$

15. $X_L = 2\pi fL = 2\pi \times 25\ kHz \times 500\ \mu H = 78.5\ \Omega$

$$\theta = \tan^{-1}\frac{X_L}{R} = \tan^{-1}\frac{78.5\ \Omega}{100\ \Omega} = 38.1°$$

$$Z = \frac{X_L}{\sin\theta} = \frac{78.5\ \Omega}{\sin 38.2°} = 127\ \Omega$$

$V_R = E\cos\theta = 12\ V \times \cos 38.2° = 9.44\ V$

$V_L = E\sin\theta = 12\ V \times \sin 38.2° = 7.41\ V$

$$I = \frac{E}{Z} = \frac{12\ V}{127\ \Omega} = 94.4\ mA$$

17. $\theta = \cos^{-1}\frac{R}{Z} = \cos^{-1}\frac{4.7\text{ k}\Omega}{6\text{ k}\Omega} = 38.4°$

$X_L = Z\sin\theta = 6\text{ k}\Omega \times \sin 38.4° = 3.73\text{ k}\Omega$

$f = \frac{X_L}{2\pi L} = \frac{3.73\text{ k}\Omega}{2\pi \times 650\text{ mH}} = 913\text{ Hz}$

$E = \frac{V_L}{\sin\theta} = \frac{7.5\text{ V}}{\sin 38.4°} = 12.1\text{ V}$

$V_R = \frac{V_L}{\tan\theta} = \frac{7.5\text{ V}}{\tan 38.4°} = 9.46\text{ V}$

$I = \frac{E}{Z} = \frac{12.1\text{ V}}{6\text{ k}\Omega} = 2.02\text{ mA}$

19. $\theta = \cos^{-1}\frac{R}{Z} = \cos^{-1}\frac{18\text{ k}\Omega}{32\text{ k}\Omega} = 55.8°$

$X_L = Z\sin\theta = 32\text{ k}\Omega \times \sin 55.8° = 26.5\text{ k}\Omega$

$f = \frac{X_L}{2\pi L} = \frac{26.5\text{ k}\Omega}{2\pi \times 75\text{ mH}} = 56.2\text{ kHz}$

$E = \frac{V_L}{\sin\theta} = \frac{3.7\text{ V}}{\sin 55.8°} = 4.47\text{ V}$

$V_R = \frac{V_L}{\tan\theta} = \frac{3.7\text{ V}}{\tan 55.8°} = 2.51\text{ V}$

$I = \frac{E}{Z} = \frac{4.47\text{ V}}{32\text{ k}\Omega} = 140\ \mu\text{A}$

21. $\theta = \cos^{-1}\frac{R}{Z} = \cos^{-1}\frac{22\text{ k}\Omega}{35.7\text{ k}\Omega} = 52°$

$X_L = Z\sin\theta = 35.7\text{ k}\Omega \times \sin 52° = 28.1\text{ k}\Omega$

$f = \frac{X_L}{2\pi L} = \frac{28.1\text{ k}\Omega}{2\pi \times 80\text{ mH}} = 5.59\text{ kHz}$

$E = \frac{V_L}{\sin\theta} = \frac{17\text{ V}}{\sin 52°} = 21.6\text{ V}$

$V_R = \frac{V_L}{\tan\theta} = \frac{17\text{ V}}{\tan 52°} = 13.3\text{V}$

$I = \frac{E}{Z} = \frac{21.6\text{ V}}{35.7\text{ k}\Omega} = 605\ \mu\text{A}$

23. $\theta = \cos^{-1}\frac{R}{Z} = \cos^{-1}\frac{750\ \Omega}{1.77\text{ k}\Omega} = 64.9°$

$X_L = Z\sin\theta = 1.77\text{ k}\Omega \times \sin 64.9° = 1.60\text{ k}\Omega$

$f = \frac{X_L}{2\pi L} = \frac{1.6\text{ k}\Omega}{2\pi \times 100\text{ mH}} = 2.55\text{ kHz}$

$$E = \frac{V_L}{\sin\theta} = \frac{27\ V}{\sin 64.9°} = 29.8\ V$$

$$V_R = \frac{V_L}{\tan\theta} = \frac{27\ V}{\tan 64.9°} = 12.6\ V$$

$$I = \frac{E}{Z} = \frac{29.8\ V}{1.77\ k\Omega} = 16.9\ mA$$

25. $$\theta = \cos^{-1}\frac{R}{Z} = \cos^{-1}\frac{6.8\ k\Omega}{11\ k\Omega} = 51.8°$$

$$X_L = Z\sin\theta = 11\ k\Omega \times \sin 51.8° = 8.65\ k\Omega$$

$$L = \frac{X_L}{2\pi f} = \frac{8.65\ k\Omega}{2\pi \times 4.3\ kHz} = 320\ mH$$

$$E = \frac{V_L}{\sin\theta} = \frac{6\ V}{\sin 51.8°} = 7.63\ V$$

$$V_R = E\cos\theta = 7.63\ V \times \cos 51.8° = 4.72\ V$$

$$I = \frac{E}{Z} = \frac{7.63\ V}{11\ k\Omega} = 694\ \mu A$$

27. $$\theta = \cos^{-1}\frac{R}{Z} = \cos^{-1}\frac{2.2\ k\Omega}{4.8\ k\Omega} = 62.7°$$

$$X_L = Z\sin\theta = 4.8\ k\Omega \times \sin 62.7° = 4.27\ k\Omega$$

$$L = \frac{X_L}{2\pi f} = \frac{4.27\ k\Omega}{2\pi \times 7.5\ kHz} = 90.5\ mH$$

$$E = \frac{V_L}{\sin\theta} = \frac{4.5\ V}{\sin 62.7°} = 5.06\ V$$

$$V_R = E\cos\theta = 5.06\ V \times \cos 62.7° = 2.32\ V$$

$$I = \frac{E}{Z} = \frac{5.06\ V}{4.8\ k\Omega} = 1.05\ mA$$

29. $$\theta = \cos^{-1}\frac{R}{Z} = \cos^{-1}\frac{47\ k\Omega}{83\ k\Omega} = 55.5°$$

$$X_L = Z\sin\theta = 83\ k\Omega \times \sin 55.5° = 68.4\ k\Omega$$

$$L = \frac{X_L}{2\pi f} = \frac{68.4\ k\Omega}{2\pi \times 5.5\ kHz} = 1.98\ H$$

$$E = \frac{V_L}{\sin\theta} = \frac{20\ V}{\sin 55.5°} = 24.3\ V$$

$$V_R = E\cos\theta = 24.3\ V \times \cos 55.5° = 13.8\ V$$

$$I = \frac{E}{Z} = \frac{24.3\ V}{83\ k\Omega} = 293\ \mu A$$

31. $\theta = \cos^{-1}\frac{R}{Z} = \cos^{-1}\frac{3.9\ \text{k}\Omega}{6.75\ \text{k}\Omega} = 54.7^\circ$

$X_L = Z\sin\theta = 6.75\ \text{k}\Omega \times \sin 54.7^\circ = 5.51\ \text{k}\Omega$

$L = \frac{X_L}{2\pi f} = \frac{5.51\ \text{k}\Omega}{2\pi \times 12\ \text{kHz}} = 73.1\ \text{mH}$

$E = \frac{V_L}{\sin\theta} = \frac{8.5\ \text{V}}{\sin 54.7^\circ} = 10.4\ \text{V}$

$V_R = E\cos\theta = 10.4\ \text{V} \times \cos 54.7^\circ = 6.01\ \text{V}$

$I = \frac{E}{Z} = \frac{10.4\ \text{V}}{6.75\ \text{k}\Omega} = 1.54\ \text{mA}$

***Problems 20-5***

1. $Z_r = 20.7\ \text{k}\Omega + j23\ \text{k}\Omega$
   $Z_p = 30.9\ \underline{/48^\circ}\ \text{k}\Omega$

3. $Z_r = 599\ \Omega - j740\ \Omega$
   $Z_p = 9526\ \underline{/51^\circ}\ \Omega$

5. $Z_p = 857\ \underline{/-17^\circ}\ \Omega$
   $Z_r = 820\ \Omega - j251\ \Omega$

7. $Z_p = 6.35\ \underline{/-45^\circ}\ \text{k}\Omega$
   $Z_r = 4.49\ \text{k}\Omega - j4.49\ \text{k}\Omega$

9. $Z_r = 40\ \text{k}\Omega - j30\ \text{k}\Omega$
   $Z_p = 50\ \underline{/-36^\circ}\ \text{k}\Omega$

11. $Z_r = 180\ \text{k}\Omega + j120\ \text{k}\Omega$
    $Z_p = 216\ \underline{/33.7^\circ}\ \text{k}\Omega$

13. $Z_p = 4.6\ \underline{/-40.7^\circ}\ \text{k}\Omega$
    $Z_r = 3.49\ \text{k}\Omega - j3\ \text{k}\Omega$

15. $Z_p = 5.35\ \underline{/44.5^\circ}\ \text{k}\Omega$
    $Z_r = 3.82\ \text{k}\Omega + j3.75\ \text{k}\Omega$

17. $Z_r = 10\ \text{k}\Omega + j18\ \text{k}\Omega$
    $Z_p = 20.6\ \underline{/61.0^\circ}\ \text{k}\Omega$

19. $Z_r = 1.8\ \text{k}\Omega - j1.45\ \text{k}\Omega$
    $Z_p = 2.31\ \underline{/-38.9^\circ}\ \text{k}\Omega$

21. $E_r = 10\ \text{V} + j12.7\ \text{V}$
    $E_p = 16.2\ \underline{/51.8^\circ}\ \text{V}$

23. $E_r = 4.32\ \text{V} - j6.25\ \text{V}$
    $E_p = 7.60\ \underline{/-55.4^\circ}\ \text{V}$

25. $E_p = 12\ \underline{/-28.4^\circ}\ \text{V}$
    $E_r = 10.6\ \text{V} - j5.7\ \text{V}$

27. $E_p = 15\ \underline{/35.5^\circ}\ \text{V}$
    $E_r = 12.4\ \text{V} + j8.5\ \text{V}$

29. $E_p = 15\ \underline{/-40^\circ}\ \text{V}$
    $E_r = 11.5\ \text{V} - j9.64\ \text{V}$

31. $E_p = 14\ \underline{/34.6^\circ}\ \text{V}$
    $E_r = 11.5\ \text{V} + j7.93\ \text{V}$

33. $E_p = 136\ \underline{/-28^\circ}\ \text{V}$
    $E_r = 120\ \text{V} - j64\ \text{V}$

35. $E_p = 5\ \underline{/58.7^\circ}\ \text{V}$
    $E_r = 2.6\ \text{V} + j4.27\ \text{V}$

37. $E_p = 30\ \underline{/32.2^\circ}\ \text{V}$
    $E_r = 25.4\ \text{V} + j16\ \text{V}$

39. $E_p = 10\ \underline{/-36.9^\circ}\ \text{V}$
    $E_r = 8\ \text{V} - j6\ \text{V}$

***Problems 20-6***

1. $Z_{1(r)} = 1.88\ \text{k}\Omega + j684\ \text{k}\Omega$
   $Z_{2(r)} = \underline{1.27\ \text{k}\Omega + j2.72\ \text{k}\Omega}$
   $Z_{T(r)} = 3.15\ \text{k}\Omega + 3.40\ \text{k}\Omega$
   $Z_{T(p)} = 4.64\ \underline{/47.2^\circ}\ \text{k}\Omega$

3. $Z_{1(r)} = 10.3\ \text{k}\Omega - j3.76\ \text{k}\Omega$
   $Z_{2(r)} = \underline{13\ \text{k}\Omega - j7.5\ \text{k}\Omega}$
   $Z_{T(r)} = 23.3\ \text{k}\Omega - j11.3\ \text{k}\Omega$
   $Z_{T(p)} = 25.9\ \underline{/-25.8^\circ}\ \text{k}\Omega$

5. $Z_{1(r)} = 6.62\ k\Omega - j3.09\ k\Omega$
$Z_{2(r)} = \underline{2.3\ k\Omega + j1.93\ k\Omega}$
$Z_{T(r)} = 8.92\ k\Omega - j1.16\ k\Omega$
$Z_{T(p)} = 9.00\ \underline{/-7.41^\circ}\ k\Omega$

7. $Z_{1(r)} = 61.5\ k\Omega + j48.0\ k\Omega$
$Z_{2(r)} = \underline{21.1\ k\Omega - j45.3\ k\Omega}$
$Z_{T(r)} = 82.6\ k\Omega + j2.7\ k\Omega$
$Z_{T(p)} = 82.6\ \underline{/1.87^\circ}\ k\Omega$

9. $Z_{1(r)} = 234\ k\Omega + j135\ k\Omega$
$Z_{2(r)} = \underline{130\ k\Omega + j75\ k\Omega}$
$Z_{T(r)} = 364\ k\Omega + j210\ k\Omega$
$Z_{T(p)} = 420\ \underline{/30^\circ}\ k\Omega$

11. $Z_{1(r)} = 0 - j7.5\ k\Omega$
$Z_{2(r)} = \underline{12.5\ k\Omega + j5.06\ k\Omega}$
$Z_{T(r)} = 12.5\ k\Omega - j2.44\ k\Omega$
$Z_{T(p)} = 12.8\ \underline{/-11.0^\circ}\ k\Omega$

***Problems 20-7***

1. $Z_r = 1.76\ k\Omega + j3\ k\Omega$
$Z_p = 3.48\ \underline{/59.6^\circ}\ k\Omega$

3. $Z_r = 1.15\ k\Omega + j1.3\ k\Omega$
$Z_p = 1.74\ \underline{/48.5^\circ}\ k\Omega$

5. $X_C = \dfrac{1}{2\pi fC} = \dfrac{1}{2\pi \times 1.2\ kHz \times 100\ nF} = 1.33\ k\Omega$

$X_L = 2\pi fL = 2\pi \times 1.2\ kHz \times 800\ mH = 6.03\ k\Omega$

$Z_r = 3.4\ k\Omega + j4.70\ k\Omega$ $\qquad Z_p = 5.80\ \underline{/54.1^\circ}\ k\Omega$

In the ESC: $R = 3.4\ k\Omega$ and $L = \dfrac{X_L}{2\pi f} = \dfrac{4.7\ k\Omega}{2\pi \times 1.2\ kHz} = 623\ mH$

7. $X_C = \dfrac{1}{2\pi fC} = \dfrac{1}{2\pi \times 500\ Hz \times 350\ nF} = 909\ \Omega$

$X_L = 2\pi fL = 2\pi \times 500\ Hz \times 200\ mH = 628\ \Omega$

$Z_r = 200\ \Omega - j281\ \Omega$ $\qquad Z_p = 345\ \underline{/-54.5^\circ}\ \Omega$

In the ESC: $R = 200\ \Omega$ and $C = \dfrac{1}{2\pi fX_C} = \dfrac{1}{2\pi \times 500\ Hz \times 281\ \Omega} = 1.13\ \mu F$

$I = \dfrac{E}{Z} = \dfrac{25\ V}{345\ \Omega} = 72.5\ mA$

$V_R = IR = 72.5\ mA \times 200\ \Omega = 14.5\ V$

$V_C = IX_C = 72.5\ mA \times 909\ \Omega = -65.9\ V$

$V_L = IX_L = 72.5\ mA \times 628\ \Omega = 45.5\ V$

***Problems 20-8***

1. $f_r = \dfrac{1}{2\pi\ [LC]^{1/2}} = \dfrac{1}{2\pi \times [750\ mH \times 5\ nF]^{1/2}} = 2.60\ kHz$

at $f_r$: $X_C = \dfrac{1}{2\pi fC} = \dfrac{1}{2\pi \times 2.60\ kHz \times 5\ nF} = 12.2\ k\Omega$

$X_L = 2\pi fL = 2\pi \times 2.60\ kHz \times 750\ mH = 12.2\ k\Omega$

$Z_r = 470\ \Omega + j12.2\ k\Omega - j12.2\ k\Omega = 470\ \Omega + j0$

$Z_p = 470\ \underline{/0^\circ}\ \Omega$

$I = \dfrac{E}{Z} = \dfrac{10\ V}{470\ \Omega} = 21.3\ mA$

$V_R = IR = 21.3\ mA \times 470\ \Omega = 10\ V$

$V_C = IX_C = 21.3\ mA \times 12.2\ k\Omega = -260\ V$

$V_L = IX_L = 21.3\ mA \times 12.2\ k\Omega = 260\ V$

The ESC is: $R = 470\ \Omega$

at $f = 2.3$ kHz:

$$X_C = \frac{1}{2\pi fC} = \frac{1}{2\pi \times 2.30 \text{ kHz} \times 5 \text{ nF}} = 13.8 \text{ k}\Omega$$

$X_L = 2\pi fL = 2\pi \times 2.30 \text{ kHz} \times 750 \text{ mH} = 10.8 \text{ k}\Omega$

$Z_r = 470\ \Omega + j10.8 \text{ k}\Omega - j13.8 \text{ k}\Omega = 470\ \Omega - j3.00 \text{ k}\Omega$

$Z_p = 3.04\ \angle{-81.1°}\ \text{k}\Omega$

$$I = \frac{E}{Z} = \frac{10 \text{ V}}{3.04 \text{ k}\Omega} = 3.29 \text{ mA}$$

$V_R = IR = 3.29 \text{ mA} \times 470\ \Omega = 1.55 \text{ V}$

$V_C = IX_C = 3.29 \text{ mA} \times 13.8 \text{ k}\Omega = -45.4 \text{ V}$

$V_L = IX_L = 3.29 \text{ mA} \times 10.8 \text{ k}\Omega = 35.5 \text{ V}$

The ESC is:

$$R = 470\ \Omega \text{ and } C = \frac{1}{2\pi \times 2.3 \text{ kHz} \times 3.04 \text{ k}\Omega} = 23.1 \text{ nF}$$

at $f = 2.7$ kHz:

$$X_C = \frac{1}{2\pi fC} = \frac{1}{2\pi \times 2.70 \text{ kHz} \times 5 \text{ nF}} = 11.8 \text{ k}\Omega$$

$X_L = 2\pi fL = 2\pi \times 2.70 \text{ kHz} \times 750 \text{ mH} = 12.7 \text{ k}\Omega$

$Z_r = 470\ \Omega + j12.7 \text{ k}\Omega - j11.8 \text{ k}\Omega = 470\ \Omega + j900\ \Omega$

$Z_p = 1.02\ \angle 62.4°\ \text{k}\Omega$

$$I = \frac{E}{Z} = \frac{10 \text{ V}}{1.02 \text{ k}\Omega} = 9.80 \text{ mA}$$

$V_R = IR = 9.80 \text{ mA} \times 470\ \Omega = 4.61 \text{ V}$

$V_C = IX_C = 9.80 \text{ mA} \times 11.8 \text{ k}\Omega = -116 \text{ V}$

$V_L = IX_L = 9.80 \text{ mA} \times 12.7 \text{ k}\Omega = 125 \text{ V}$

The ESC is:

$$R = 470\ \Omega \text{ and } L = \frac{900\ \Omega}{2\pi \times 2.7 \text{ kHz}} = 53.1 \text{ mH}$$

3. $$f_r = \frac{1}{2\pi\ [LC]^{1/2}} = \frac{1}{2\pi \times [1 \text{ H} \times 500 \text{ nF}]^{1/2}} = 225 \text{ Hz}$$

at $f_r$: $X_L = 2\pi fL = 2\pi \times 225 \text{ Hz} \times 1 \text{ H} = 1.41 \text{ k}\Omega$

$$X_C = \frac{1}{2\pi fC} = \frac{1}{2\pi \times 225 \text{ Hz} \times 500 \text{ nF}} = 1.41 \text{ k}\Omega$$

$Z_r = 300\ \Omega + j1.41 \text{ k}\Omega - j1.41 \text{ k}\Omega = 300\ \Omega + j0$

$Z_p = 300\ \angle 0°\ \Omega$

$$I = \frac{E}{Z} = \frac{25 \text{ V}}{300\ \Omega} = 83.3 \text{ mA}$$

$V_R = IR = 83.3 \text{ mA} \times 300\ \Omega = 25 \text{ V}$

$V_C = IX_C = 83.3 \text{ mA} \times 1.41 \text{ k}\Omega = -118 \text{ V}$

$V_L = IX_L = 83.3 \text{ mA} \times 1.41 \text{ k}\Omega = 118 \text{ V}$

The ESC is: $R = 300\ \Omega$

at $f = 200$ Hz:

$$X_C = \frac{1}{2\pi fC} = \frac{1}{2\pi \times 200 \text{ Hz} \times 500 \text{ nF}} = 1.59 \text{ k}\Omega$$

$X_L = 2\pi fL = 2\pi \times 200 \text{ Hz} \times 1 \text{ H} = 1.26 \text{ k}\Omega$

$Z_r = 300\ \Omega + j1.26\ k\Omega - j1.59\ k\Omega = 300\ \Omega - j330\ \Omega$

$Z_p = 446\ \angle -47.7°\ \Omega$

$I = \frac{E}{Z} = \frac{25\ V}{446\ \Omega} = 56.1\ mA$

$V_R = IR = 56.1\ mA \times 300\ \Omega = 16.8\ V$

$V_C = IX_C = 56.1\ mA \times 1.59\ k\Omega = -89.1\ V$

$V_L = IX_L = 56.1\ mA \times 1.26\ k\Omega = 70.6\ V$

The ESC is:

$R = 300\ \Omega$ and $C = \frac{1}{2\pi \times 200\ Hz \times 330\ \Omega} = 2.41\ \mu F$

at f = 300 Hz

$X_C = \frac{1}{2\pi fC} = \frac{1}{2\pi \times 300\ Hz \times 500\ nF} = 1.06\ k\Omega$

$X_L = 2\pi fL = 2\pi \times 300\ Hz \times 1\ H = 1.88\ k\Omega$

$Z_r = 300\ \Omega + j1.88\ k\Omega - j1.06\ k\Omega = 300\ \Omega + j819\ \Omega$

$Z_p = 872\ \angle 69.9°\ \Omega$

$I = \frac{E}{Z} = \frac{25\ V}{872\ \Omega} = 28.7\ mA$

$V_R = IR = 28.7\ mA \times 300\ \Omega = 8.60\ V$

$V_C = IX_C = 28.7\ mA \times 1.06\ k\Omega = -30.4\ V$

$V_L = IX_L = 28.7\ mA \times 1.88\ k\Omega = 53.9\ V$

The ESC is:

$R = 300\ \Omega$ and $L = \frac{819\ \Omega}{2\pi \times 300\ Hz} = 434\ mH$

# CHAPTER 21

***Problems 21-1***

1. $I_P = 40 \angle 54^\circ$

   $I_R = 23.5\text{ mA} + j32.4\text{ mA}$

3. $I_P = 2.75 \angle 76.4^\circ\text{mA}$

   $I_R = 1.05\text{ mA} + j2.54\text{ mA}$

5. $I_P = 175 \angle 22^\circ\ \mu\text{A}$

   $I_R = 162\ \mu\text{A} + j65.6\ \mu\text{A}$

7. $I_P = 442 \angle 37.6^\circ\ \mu\text{A}$

   $I_R = 350\ \mu\text{A} + j270\ \mu\text{A}$

9. $I_P = 8.51 \angle 55.9^\circ\text{ mA}$

   $I_R = 4.77\text{ mA x } j7.05\text{ mA}$

11. $I_P = 141 \angle 45^\circ\ \mu\text{A}$

    $I_R = 100\ \mu\text{A} + j100\ \mu\text{A}$

13. $I_P = 1.19 \angle 33^\circ\text{mA}$

    $I_R = 1\text{ mA} + j649\text{ mA}$

15. $I_P = 683 \angle 20.5^\circ\ \mu\text{A}$

    $I_R = 640\ \mu\text{A} + j239\ \mu\text{A}$

17. $I_P = \angle 53.9^\circ\text{ mA}$

    $I_R = 25.3\text{ mA} + j47.6\text{ mA}$

19. $I_P = 47.6 \angle 71^\circ\text{ mA}$

    $I_R = 15.5\text{ mA} + j45\text{ mA}$

21. $I_P = 6.31 \angle 59.5^\circ\text{ mA}$

    $I_R = 3.2\text{ mA} + j5.44\text{ mA}$

23. $I_P = 333 \angle 55.7^\circ\ \mu\text{A}$

    $I_R = 188\ \mu\text{A} + j275\ \mu\text{A}$

25. $I_P = 12 \angle 51.3^\circ\text{ mA}$

    $I_R = 7.5\text{ mA} + j9.37\text{ mA}$

27. $I_P = 1 \angle 49.5^\circ\text{ mA}$

    $I_R = 650\ \mu\text{A} + j760\ \mu\text{A}$

29. $I_P = 3.15 \angle 36.0^\circ\text{ mA}$

    $I_R = 2.55\text{ mA} + j1.85\text{ mA}$

31. $I_P = 43 \angle 35.5^\circ\text{ mA}$

    $I_R = 35\text{ mA} + j25\text{mA}$

33. $I_P = 400 \angle 30^\circ\ \mu\text{A}$

    $I_R = 346\ \mu\text{A} + j200\ \mu\text{A}$

35. $I_P = 6.08 \angle 35.7^\circ\text{ mA}$

    $I_R = 4.94\text{ mA} + j3.55\text{ mA}$

***Problems 21-2***

1. $B_C = 2\pi fC = 2\pi \times 2\text{ kHz} \times 30\text{ nF} = 377\ \mu\text{S}$

   $G = \frac{1}{R} = \frac{1}{3.3\text{ k}\Omega} = 303\ \mu\text{S}$

   $Y_r = 303\ \mu\text{S} + j377\ \mu\text{S}$

   $Y_P = 484 \angle 51.2^\circ\ \mu\text{S}$

   $I_r = 2.51\text{ mA} + j3.12\text{ mA}$

   $V = \frac{I_T}{Y} = \frac{4\text{ mA}}{484\ \mu\text{S}} = 8.26\text{ V}$

3. $B_C = 2\pi fC = 2\pi \times 60\text{ kHz} \times 5\text{ nF} = 1.88\text{ mS}$

   $G = \frac{1}{R} = \frac{1}{750\ \Omega} = 1.33\text{ mS}$

   $Y_r = 1.33\text{ mS} + j1.88\text{ mS}$

   $Y_P = 2.31 \angle 54.7^\circ\text{ mS}$

   $I_r = 577\ \mu\text{A} + j816\ \mu\text{A}$

   $V = \frac{I_T}{Y} = \frac{1\text{ mA}}{2.31\text{ mS}} = 433\text{ mV}$

5. $B_C = 2\pi fC = 2\pi \times 1.5\text{ kHz} \times 10\text{ nF} = 94.2\ \mu S$

$G = \frac{1}{R} = \frac{1}{6.8\text{ k}\Omega} = 147\ \mu S$

$Y_r = 147\ \mu S + j94.2\ \mu S$

$Y_P = 175\ \underline{/32.7°}\ \mu S$

$I_r = 16.8\text{ mA} + j10.8\text{ mA}$

$V = \frac{I_T}{Y} = 115\text{ V}$

7. $B_C = 2\pi fC = 2\pi \times 1\text{ kHz} \times 30\text{ nF} = 188\ \mu S$

$G = \frac{1}{R} = \frac{1}{10\text{ k}\Omega} = 100\ \mu S$

$Y_r = 100\ \mu S + j188\ \mu S$

$Y_P = 213\ \underline{/62.1°}\ \mu S$

$I_r = 4.69\text{ mA} + j8.83\text{ mA}$

$V = \frac{I_T}{Y} = \frac{10\text{ mA}}{213\mu S} = 46.9\text{V}$

9. $B_C = 2\pi fC = 2\pi \times 35\text{ kHz} \times 300\text{ pF} = 66.0\ \mu S$

$G = \frac{1}{R} = \frac{1}{15\text{ k}\Omega} = 66.7\ \mu S$

$Y_r = 66.7\ \mu S + j66.0\ \mu S$

$Y_P = 93.8\ \underline{/44.7°}\ \mu S$

$I_r = 1.42\text{ mA} + j1.41\text{ mA}$

$V = \frac{I_T}{Y} = 21.3\text{ V}$

11. $G = \frac{B_C}{\tan\theta} = \frac{200\ \mu S}{\tan 50°} = 168\ \mu S$

$R = \frac{1}{G} = \frac{1}{168\ \mu S} = 5.96\text{ k}\Omega$

$f = \frac{B_C}{2\pi C} = \frac{200\ \mu S}{2\pi \times 25\text{ nF}} = 1.27\text{ kHz}$

$I_T = \frac{I_R}{\cos\theta} = \frac{3\text{ mA}}{\cos 50°} = 4.67\text{ mA}$

$V = I_R R = 3\text{ mA} \times 5.96\text{ k}\Omega = 17.9\text{ V}$

***Problems 21-3***

1. $I_P = 3.8\ \underline{/-30°}\text{ mA}$

$I_R = 3.29\text{ mA} - j1.9\text{ mA}$

3. $I_P = 210\ \underline{/-39°}\ \mu A$

$I_R = 163\ \mu A - j132\ \mu A$

5. $I_P = 1.34\ \underline{/-23.3°}\text{ mA}$

$I_R = 1.23\text{ mA} - j532\ \mu A$

7. $I_P = 41.8\ \underline{/-73.3°}\text{ mA}$

$I_R = 12\text{ mA} - j40\text{ mA}$

9. $I_P = 50\ \underline{/-53.1°}\text{ mA}$

$I_R = 30\text{ mA} -j40\text{ mA}$

11. $I_P = 234\ \underline{/-50.2°}\ \mu A$

$I_R = 150\ \mu A -j180\ \mu A$

13. $I_P = 2.28\ \underline{/-28.5°}\text{mA}$

$I_R = 2\text{ mA} -j1.09\text{ mA}$

15. $I_P = 4.45\ \underline{/-73°}\text{ mA}$

$I_R = 1.3\text{ mA} -j4.25\text{ mA}$

17. $I_P = 42.4\ \underline{/-27.4°}\ \mu A$

$I_R = 37.6\ \mu A -j19.5\ \mu A$

19. $I_P = 1.03\ \underline{/-45°}\ \mu A$

$I_R = 730\ \mu A -j730\ \mu A$

21. $I_P = 14\ \underline{/-40°}\text{ mA}$

$I_R = 10.7\text{ mA} -j9\text{ mA}$

23. $I_P = 690\ \underline{/-43.5°}\ \mu A$

$I_R = 501\ \mu A -j475\ \mu A$

25. $I_P = 27 \underline{/-47.8°}$ mA

$I_R$ = 18.1 mA -j20 mA

27. $I_P = 300 \underline{/-30°}$ μA

$I_R$ = 260 μA -j150 μA

29. $I_P = 1.83 \underline{/-30.4°}$ mA

$I_R$ = 1.58 mA -j925 μA

31. $I_P = 6.7 \underline{/-45.5°}$ mA

$I_R$ = 4.7 mA -j4.77 mA

33. $I_P = 1.35 \underline{/-52.3°}$ mA

$I_R$ = 825 μA -j1.07 mA

35. $I_P = 100 \underline{/-42.4°}$ mA

$I_R$ = 73.8 mA -j76.5 mA

***Problems 21-4***

1. $B_L = \frac{1}{2\pi fL} = \frac{1}{2\pi \times 200 \text{ kHz} \times 20 \text{ mH}} = 39.8\ \mu S$

$G = \frac{1}{R} = \frac{1}{20\text{ k}\Omega} = 50\ \mu S$

$Y_r = 50\ \mu S - j39.8\ \mu S$

$Y_p = 63.9 \underline{/-38.5°}\ \mu S$

$I_r$ = 31.3 mA - j24.9 mA

$I_p = 40 \underline{/-38.5°}$ mA

3. $B_L = \frac{1}{2\pi fL} = \frac{1}{2\pi \times 20 \text{ kHz} \times 60 \text{ mH}} = 133\ \mu S$

$G = \frac{1}{R} = \frac{1}{10\text{ k}\Omega} = 100\ \mu S$

$Y_r = 100\ \mu S - j133\ \mu S$

$Y_p = 166 \underline{/-53°}\ \mu S$

$I_r$ = 120 mA - j160 mA

$I_p = 200 \underline{/-53°}\ \mu S$

5. $B_L = \frac{1}{2\pi fL} = \frac{1}{2\pi \times 10 \text{ kHz} \times 100 \text{ mH}} = 159\ \mu S$

$G = \frac{1}{R} = \frac{1}{4.7\text{ k}\Omega} = 213\ \mu S$

$Y_r = 213\ \mu S - j159\ \mu S$

$Y_p = 266 \underline{/-36.8°}\ \mu S$

$I_r$ = 80.1 μA - j59.9 μA

$I_p = 100 \underline{/-36.8°}\ \mu A$

7. $B_L = \frac{1}{2\pi fL} = \frac{1}{2\pi \times 2 \text{ kHz} \times 2.5 \text{ H}} = 31.8\ \mu S$

$G = \frac{1}{R} = \frac{1}{27\text{ k}\Omega} = 37.0\ \mu S$

$Y_r = 37\ \mu S - j31.9\ \mu S$

$Y_p = 48.8 \underline{/-40.8°}\ \mu S$

$I_r$ = 303 μA - j261 μA

$I_p = 400 \underline{/-40.8°}\ \mu A$

9. $B_L = \frac{1}{2\pi fL} = \frac{1}{2\pi \times 400 \text{ Hz} \times 600 \text{ mH}} = 663\ \mu S$

$G = \frac{1}{R} = \frac{1}{750\ \Omega} = 1.33$ mS

$Y_r = 1.33\text{ mS} - j663\ \mu S$

$Y_p = 1.49 \underline{/-26.4°}$ mS

$I_r$ = 26.9 mA - j13.4 mA

$I_p = 30 \underline{/-26.4°}$ mA

***Problems 21-5***

1. $Z_T = 269 \underline{/-27°}\ \Omega$

3. $Y_T = 158 \underline{/40°}\ \mu S$

5. $Z_T = \frac{Z_1 Z_2}{Z_1 + Z_2} = \frac{700 \underline{/70°} \times 1.5 \underline{/-25°}\text{ k}\Omega}{1.60 \underline{/0.855°}\text{ k}\Omega} = 657 \underline{/44.1°}\ \Omega$

*(See Example 21-10 for the algorithm used to add $Z_1$ and $Z_2$)*

7. $Z_T = \dfrac{Z_1 Z_2}{Z_1 + Z_2} = \dfrac{35\ \underline{/20^\circ}\ k\Omega \times 10\ \underline{/-40^\circ}\ k\Omega}{40.9\ \underline{/7.78^\circ}\ k\Omega} = 8.55\ \underline{/-27.8^\circ}\ k\Omega$

9. $Z_T = \dfrac{Z_1 Z_2}{Z_1 + Z_2} = \dfrac{3.3\ \underline{/0^\circ}\ k\Omega \times 4.2\ \underline{/-90^\circ}\ k\Omega}{5.34\ \underline{/-51.8^\circ}\ k\Omega} = 2.59\ \underline{/-38.2^\circ}\ k\Omega$

***Problems 21-6***

1. $B_C = 2\pi fC = 2\pi \times 35\ kHz \times 400\ PF = 88.0\ \mu S$

$G = \dfrac{1}{R} = \dfrac{1}{15\ k\Omega} = 66.7\ \mu S$

$Y_r = 66.7\ \mu S - j88.0\ \mu S$

$Y_p = 110\ \underline{/52.8^\circ}\ \mu S$

$Z_p = \dfrac{1}{Y_p} = 9.06\ \underline{/-52.8^\circ}\ k\Omega$

$Z_r = 5.47\ k\Omega - j7.22\ k\Omega$

$I_R = 1.21\ mA$

$I_C = 1.59\ mA$

$V = \dfrac{I_T}{Y} = \dfrac{2\ mA}{110\ \mu S} = 18.1\ V$

*In the ESC:* $R = 5.47\ k\Omega$

$C = \dfrac{1}{2\pi fX_C} = \dfrac{1}{2\pi \times 35\ kHz \times 7.22\ k\Omega} = 630\ pF$

3. $B_C = 2\pi fC = 2\pi \times 3\ kHz \times 150\ nF = 2.83\ mS$

$G = \dfrac{1}{R} = \dfrac{1}{470\ \Omega} = 2.13\ mS$

$Y_r = 2.13\ mS + j2.83\ mS$

$Y_p = 3.54\ \underline{/53.0^\circ}\ mS$

$Z_p = \dfrac{1}{Y_p} = 283\ \underline{/-53^\circ}\ \Omega$

$Z_r = 170\ \Omega - j226\ \Omega$

$I_R = 24.1\ mA$

$I_L = 32.0\ mA$

$V = \dfrac{I_T}{Y} = \dfrac{40\ mA}{3.54\ \mu S} = 11.3\ V$

*In the ESC:* $R = 170\ \Omega$

$C = \dfrac{1}{2\pi fX_C} = \dfrac{1}{2\pi \times 3\ kHz \times 226\ \Omega} = 236\ nF$

5. $B_L = \dfrac{1}{2\pi fL} = \dfrac{1}{2\pi \times 150\ kHz \times 60\ mH} = 17.7\ \mu S$

$G = \dfrac{1}{R} = \dfrac{1}{47\ k\Omega} = 21.3\ \mu S$

$Y_r = 21.3\ \mu S - j17.7\ \mu S$

$Y_p = 27.7\ \underline{/-39.7^\circ}\ \mu S$

$Z_p = 36.1\ \underline{/39.7^\circ}\ k\Omega$

$Z_r = 27.8\ k\Omega + j23.1\ k\Omega$

$I_R = 192\ \mu A$

$I_L = 160\ \mu A$

$V = \dfrac{I_T}{Y} = \dfrac{250\ \mu A}{27.7\ \mu S} = 9.04\ V$

*In the ESC:* $R = 27.8\ k\Omega$

$$L = \frac{X_L}{2\pi f} = \frac{23.1\ k\Omega}{2\pi \times 150\ kHz} = 24.5\ mH$$

7. $B_L = \frac{1}{2\pi fL} = \frac{1}{2\pi \times 5\ kHz \times 30\ mH} = 1.06\ mS$

$G = \frac{1}{R} = \frac{1}{680\ \Omega} = 1.47\ mS$

$Y_r = 1.47\ mS - j1.06\ mS$

$Y_p = 1.81\ \underline{/35.8°}\ mS$

$Z_p = 552\ \underline{/35.8°}\ \Omega$

$I_R = 4.06\ mA$

$I_L = 2.92\ mA$

$V = \frac{I_T}{Y} = \frac{5\ mA}{1.81\ mS} = 2.76\ V$

$Z_r = 447\ \Omega + j323\ \Omega$

*In the ESC:* $R = 447\ \Omega$

$$L = \frac{X_L}{2\pi f} = \frac{323\ \Omega}{2\pi \times 5\ kHz} = 10.3\ mH$$

9. $B_L = \frac{1}{2\pi fL} = \frac{1}{2\pi \times 25\ kHz \times 30\ mH} = -212\ \mu S$

$B_C = 2\pi fC = 2\pi \times 25\ kHz \times 300\ pF = 47.1\ \mu S$

$G = \frac{1}{R} = \frac{1}{10\ k\Omega} = 100\ \mu S$

$Y_r = 100\ \mu S - j165\ \mu S$

$Y_p = 193\ \underline{/-58.8°}\ \mu S$

$Z_p = 5.18\ \underline{/58.8°}\ k\Omega$

$I_R = VG = 104\ V \times 100\ \mu S = 10.4\ mA$

$I_C = VB_C = 104\ V \times 47.1\ \mu S = 4.88\ mA$

$V = \frac{I_T}{Y} = \frac{20\ mA}{193\ \mu S} = 104\ V$

$Z_r = 2.68\ k\Omega + j4.43\ k\Omega$

$I_L = VB_L = 104\ V \times -212\ \mu S = -22.0\ mA$

*In the ESC:* $R = 2.68\ k\Omega$

$$L = \frac{X_L}{2\pi f} = \frac{4.43\ k\Omega}{2\pi \times 25\ kHz} = 28.2\ mH$$

11. $B_L = \frac{1}{2\pi fL} = \frac{1}{2\pi \times 15\ kHz \times 4\ mH} = 2.65\ mS$

$B_C = 2\pi fC = 2\pi \times 15\ kHz \times 30\ nF = 2.83\ mS$

$G = \frac{1}{R} = \frac{1}{6.8\ k\Omega} = 147\ \mu S$

$Y_r = 147\ \mu S + j175\ \mu S$

$Y_p = 228\ \underline{/49.9°}\ \mu S$

$I_R = VG = 1.75\ V \times 147\ \mu S = 257\ \mu A$

$I_C = VB_C = 1.75\ V \times 2.83\ mS = 4.95\ mA$

$Z_p = 4.38\ \underline{/-49.9°}\ k\Omega$

$Z_r = 2.82\ k\Omega - j3.35\ k\Omega$

$$V = \frac{I_T}{Y} = \frac{400\ \mu A}{228\ \mu S} = 1.75\ V$$

$$I_L = VB_L = 1.75\ V \times 2.65\ mS = -4.64\ mA$$

*In the ESC:* $R = 2.82\ k\Omega$

$$C = \frac{1}{2\pi f X_C} = \frac{1}{2\pi \times 15\ kHz \times 3.35\ k\Omega} = 3.17\ nF$$

***Problems 21-7***

1. $$X_C = \frac{1}{2\pi fC} = \frac{1}{2\pi \times 4\ kHz \times 50\ nF} = 796\ \Omega$$

$$G = \frac{R}{R^2 + X_C^2} = \frac{680\ \Omega}{680\ \Omega^2 + 796\ \Omega^2} = 621\ \mu S$$

$$B_C = \frac{X_C}{R^2 + X_C^2} = \frac{796\ \Omega}{680\ \Omega^2 + 796\ \Omega^2} = 726\ \mu S$$

$$Y_r = 621\ \mu S + j726\ \mu S$$

$$C = \frac{B_C}{2\pi f} = \frac{726\ \mu S}{2\pi \times 4\ kHz} = 28.9\ nF$$

3. $$X_C = \frac{1}{2\pi fC} = \frac{1}{2\pi \times 120\ kHz \times 100\ pF} = 13.3\ k\Omega$$

$$G = \frac{R}{R^2 + X_C^2} = \frac{15\ k\Omega}{15\ k\Omega^2 + 13.3\ k\Omega^2} = 37.4\ \mu S$$

$$B_C = \frac{X_C}{R^2 + X_C^2} = \frac{13.3\ k\Omega}{15\ k\Omega^2 + 13.3\ k\Omega^2} = 33.1\ \mu S$$

$$Y_r = 37.4\ \mu S + j33.1\ \mu S$$

$$C = \frac{B_C}{2\pi f} = \frac{33.1\ \mu S}{2\pi \times 120\ kHz} = 43.9\ pF$$

5. $$X_L = 2\pi fL = 2\pi \times 2.65\ kHz \times 100\ mH = 1.67\ k\Omega$$

$$G = \frac{R}{R^2 + X_L^2} = \frac{2.2\ k\Omega}{2.2\ k\Omega^2 + 1.67\ k\Omega^2} = 289\ \mu S$$

$$B_L = \frac{X_L}{R^2 + X_L^2} = \frac{1.67\ k\Omega}{2.2\ k\Omega^2 + 1.67\ k\Omega^2} = 219\ \mu S$$

$$Y_r = 289\ \mu S - j219\ \mu S$$

$$L = \frac{1}{2\pi f B_L} = \frac{1}{2\pi \times 2.65\ k \times 219\ \mu S} = 274\ mH$$

7. $X_L = 2\pi fL = 2\pi \times 10\text{ kHz} \times 200\text{ mH} = 12.6\text{ k}\Omega$

$$G = \frac{R}{R^2 + X_L^2} = \frac{15\text{ k}\Omega}{15\text{ k}\Omega^2 + 12.6\text{ k}\Omega^2} = 39.2\ \mu S$$

$$B_L = \frac{X_L}{R^2 + X_L^2} = \frac{12.6\text{ k}\Omega}{15\text{ k}\Omega^2 + 12.6\text{ k}\Omega^2} = 32.8\ \mu S$$

$Y_r = 39.2\ \mu S - j32.8\ \mu S$

$$L = \frac{1}{2\pi fB_L} = \frac{1}{2\pi \times 10\text{ kHz} \times 32.8\ \mu S} = 485\text{ mH}$$

9. $B_C = 2\pi fC = 2\pi \times 10\text{ kHz} \times 2\text{ nF} = 126\ \mu S$

$$R = \frac{G}{G^2 + B_C^2} = \frac{256\ \mu S}{256\ \mu S^2 + 126\ \mu S^2} = 3.15\text{ k}\Omega$$

$$X_C = \frac{B_C}{G^2 + B_C^2} = \frac{126\ \mu S}{256\ \mu S^2 + 126\ \mu S^2} = 1.55\text{ k}\Omega$$

$Z_r = 3.15\text{ k}\Omega - j1.55\text{ k}\Omega$

$$C = \frac{1}{2\pi fX_C} = \frac{1}{2\pi \times 10\text{ kHz} \times 1.55\text{ k}\Omega} = 10.3\text{ nF}$$

11. $B_C = 2\pi fC = 2\pi \times 3.25\text{ kHz} \times 100\text{ nF} = 2.04\text{ mS}$

$$R = \frac{G}{G^2 + B_C^2} = \frac{1.22\text{ mS}}{1.22\text{ mS}^2 + 2.04\text{ mS}^2} = 216\ \Omega$$

$$X_C = \frac{B_C}{G^2 + B_C^2} = \frac{2.04\text{ mS}}{1.22\text{ mS}^2 + 2.04\text{ mS}^2} = 361\ \Omega$$

$Z_r = 216\ \Omega - j361\ \Omega$

$$C = \frac{1}{2\pi fX_C} = \frac{1}{2\pi \times 3.25\text{ kHz} \times 361\ \Omega} = 136\text{ nF}$$

13. $$B_L = \frac{1}{2\pi fC} = \frac{1}{2\pi \times 100\text{ Hz} \times 200\text{ mH}} = 7.96\text{ mS}$$

$$R = \frac{G}{G^2 + B_L^2} = \frac{14.7\text{ mS}}{14.7\text{ mS}^2 + 7.96\text{ mS}^2} = 52.6\ \Omega$$

$$X_L = \frac{B_L}{G^2 + B_L^2} = \frac{7.96\text{ mS}}{14.7\text{ mS}^2 + 7.96\text{ mS}^2} = 28.5\ \Omega$$

$Z_r = 52.6\ \Omega + j28.5\ \Omega$

$$L = \frac{X_L}{2\pi f} = \frac{28.5\ \Omega}{2\pi \times 100\text{ Hz}} = 45.3\text{ mH}$$

15. $B_L = \dfrac{1}{2\pi fL} = \dfrac{1}{2\pi \times 8.1 \text{ kHz} \times 30 \text{ mH}} = 655\ \mu S$

$R = \dfrac{G}{G^2 + B_L^2} = \dfrac{110\ \mu S}{110\ \mu S^2 + 655\ \mu S^2} = 249\ \Omega$

$X_L = \dfrac{B_L}{G^2 + B_L^2} = \dfrac{655\ \mu S}{110\ \mu S^2 + 655\ \mu S^2} = 1.48\ k\Omega$

$Z_r = 249\ \Omega + j1.48\ k\Omega$

$L = \dfrac{X_L}{2\pi f} = \dfrac{1.48\ k\Omega}{2\pi \times 8.1 \text{ kHz}} = 29.2 \text{ mH}$

***Problems 21-8***

1. Convert the RL series circuit to its parallel equivalent:

   $G = 4.99\ \mu S$ and $B_L = 99.8\ \mu S$

   $Y_r = G + jB_C - jB_L = 4.99\ \mu S + j50\ \mu S - j99.8\ \mu S$

   $= 4.99\ \mu S - j49.8\ \mu S$ (equivalent circuit is RL)

   $Z_r = 2.00\ k\Omega + j19.9\ k\Omega$

3. Convert the RL series circuit to its parallel equivalent:

   $G = 60\ \mu S$ and $B_L = 20\ \mu S$

   Convert the RC series circuit to its parallel equivalent:

   $G = 80\ \mu S$ and $B_C = 40\ \mu S$

   $Y_r = G_1 + G_2 + jB_C - jB_L = 140\ \mu S + j20\ \mu S$ (equivalent circuit is RC)

   $Z_r = 7.00\ k\Omega - j1.00\ k\Omega$

5. Convert the RL series circuit to its parallel equivalent:

   $G = 13.8\ \mu S$ and $B_L = 17.9\ \mu S$

   $Y_r = G + jB_C - jB_L = 13.8\ \mu S + j32.1\ \mu S$ (equivalent circuit is RC)

   R series = $11.3\ k\Omega$ and $X_C = 26.3\ k\Omega$

   $Z_r = R_1 + R_S - jX_C = 16.9\ k\Omega - j26.3\ k\Omega$

7. Convert the series RL circuit to its parallel equivalent:

   $G_1 = 41.7\ \mu S$ and $B_L = 41.7\ \mu S$

   Convert the series RC circuit to its parallel equivalent:

   $G_2 = 91.7\ \mu S$ and $B_C = 27.5\ \mu S$

   $Y_r = G_1 + G_2 + jB_C - jB_L = 133\ \mu S - 14.2\ \mu S$ (equivalent circuit is RL)

   $Z_r$ (of the equivalent circuit) $= 7.41\ k\Omega + j789\ \Omega$

   $Z_r = R_1 + R_S + X_L - X_{C1} = 8.61\ k\Omega - 9.21\ k\Omega$

9. $B_C = 2\pi fC = 2\pi \times 1 \text{ kHz} \times 3 \text{ nF} = 18.8\ \mu S$

   $X_L = 2\pi fL = 2\pi \times 1 \text{ kHz} \times 500 \text{ mH} = 3.14\ k\Omega$

   Convert the series RL to its equivalent parallel circuit.

   $G = 92.0\ \mu S$ and $B_L = 289\ \mu S$

11. $X_C = \dfrac{1}{2\pi fC} = \dfrac{1}{2\pi \times 5\text{ kHz} \times 3\text{ nF}} = 10.6\text{ k}\Omega$

$X_L = 2\pi fL = 2\pi \times 5\text{ kHz} \times 250\text{ mH} = 7.85\text{ k}\Omega$

Convert the series RL circuit to its parallel equivalent:

$G = 8.07\ \mu S$ and $B_L = 127\ \mu S$

Convert the series RC circuit to its parallel equivalent:

$G = 1.78\ \mu S$ and $B_C = 94.2\ \mu S$

$Y_r = 9.85\ \mu S - j32.8\ \mu S$ (equivalent circuit is RL)

$Z_r = 8.45\text{ k}\Omega + j28.1\text{ k}\Omega$ $\quad$ $R = 8.45\text{ k}\Omega$

$\theta = 73.2°$

$L = \dfrac{X_L}{2\pi f} = \dfrac{28.1\text{ k}\Omega}{2\pi \times 5\text{ kHz}} = 895\text{ mH}$

13. Find the ESC of the series parallel circuit:

$X_{C2} = \dfrac{1}{2\pi \times 1\text{ kHz} \times 8\text{ nF}} = 19.9\text{ k}\Omega$

$X_L = 2\pi \times 1\text{ kHz} \times 3.5\text{ H} = 22.0\text{ k}\Omega$

Convert the series RL circuit to its equivalent circuit:

$G_1 = 22.6\ \mu S$ and $B_L = 24.9\ \mu S$

Convert the series RC circuit to its equivalent circuit:

$G_2 = 15.4\ \mu S$ and $B_C = 45.0\ \mu S$

$Y_r = 38.0\ \mu S + j20.0\ \mu S$ (equivalent circuit is RC)

$Z_r = 20.6\text{ k}\Omega - j10.9\text{ k}\Omega$

circuit $Z_r = 3.3\text{ k}\Omega + 20.6\text{ k}\Omega - j10.9\text{ k}\Omega - j7.97\text{ k}\Omega$

$= 23.9\text{ k}\Omega - j18.9\text{ k}\Omega$ $\quad$ $R = 23.9\text{ k}\Omega$

$\theta = -38.3°$

$C = \dfrac{1}{2\pi fX_C} = \dfrac{1}{2\pi \times 1\text{ kHz} \times 18.9\text{ k}\Omega} = 8.42\text{ nF}$

15. $X_L = 2\pi fL = 2\pi \times 6.7\text{ kHz} \times 150\text{ mH} = 6.31\text{ k}\Omega$

$B_C = 2\pi fC = 2\pi \times 6.7\text{ kHz} \times 10\text{ nF} = 421\ \mu S$

Convert the series RL circuit to its equivalent circuit

$G = 45.6\ \mu S$ and $B_L = 144\ \mu S$

$Y_r = 45.6\ \mu S - j144\ \mu S + j421\ \mu S = 45.6\ \mu S + j277\ \mu S$

$Z_r = 583\ \Omega - j3.52\text{ k}\Omega$

circuit $Z_r = 583\ \Omega + 5.6\text{ k}\Omega - j3.52\text{ k}\Omega = 6.18\text{ k}\Omega - j3.52\text{ k}\Omega$ $\quad$ $R = 6.18\text{ k}\Omega$

$\theta = -29.7°$

$C = \dfrac{1}{2\pi fX_C} = \dfrac{1}{2\pi \times 6.7\text{ kHz} \times 10\text{ nF}} = 6.75\text{ nF}$

# CHAPTER 22

***Problems 22-1***

1. $I_{max} = \dfrac{V_{in}}{R} = \dfrac{15\ V}{470\ \Omega} = 31.9\ mA,\ P_{max} = \dfrac{V_{in}^2}{R} = \dfrac{(15\ V)^2}{470\ \Omega} = 479\ mW$

   $V_{o(max)} = V_{in} = 15\ V$

   at $f_{co}$: $X_C = R = 470\ \Omega,\ Z = R - jX_C = 665\ \underline{/-45°}\ \Omega$

   $I = \dfrac{V_{in}}{Z} = \dfrac{15\ V}{665\ \Omega} = 22.6\ mA,\ P = I^2R = (22.6\ mA)^2 \times 470\ \Omega = 239\ mW$

   $V_o = IX_C = 22.6\ mA \times 470\ \Omega = 10.6\ V$

3. $I_{max} = \dfrac{V_{in}}{R} = \dfrac{300\ mV}{5.6\ k\Omega} = 53.6\ \mu A,\ P_{max} = \dfrac{V_{in}^2}{R} = \dfrac{(300\ mV)^2}{5.6\ k\Omega} = 16.1\ \mu W$

   $V_{o(max)} = V_{in} = 300\ mV$

   at $f_{co}$: $X_C = 5.6\ k\Omega,\ Z = R - jX_C = 7.92\ \underline{/-45°}\ k\Omega$

   $I = \dfrac{V_{in}}{Z} = \dfrac{300\ mV}{7.92\ k\Omega} = 37.9\ \mu A,\ P = I^2R = 8.03\ \mu W$

   $V_o = IX_C = 37.9\ \mu A \times 5.6\ k\Omega = 212\ mV$

5. $f_{co} = \dfrac{1}{2\pi RC} = \dfrac{1}{2\pi \times 1.2\ k\Omega \times 500\ nF} = 265\ Hz$

   $X_L = R = 1.2\ k\Omega,\ Z = R - jX_C = 1.70\ \underline{/-45°}\ k\Omega,\ I = \dfrac{V_{in}}{Z} = 11.8\ mA$

   $P = I^2R = (11.8\ mA)^2 \times 1.2\ k\Omega = 166\ mW,\ V_o = V_{in} \sin\theta = 14.1\ V$

7. $f_{co} = \dfrac{1}{2\pi RC} = \dfrac{1}{2\pi \times 2.2\ k\Omega \times 200\ nF} = 362\ Hz$

   $X_C = R = 2.2\ k\Omega,\ Z = R - jX_C = 3.11\ \underline{/-45°}\ k\Omega,\ I = \dfrac{V_{in}}{Z} = 12.9\ mA$

   $P = I^2R = (12.9\ mA)^2 \times 2.2\ k\Omega = 364\ mW,\ V_o = V_{in} \sin\theta = 28.3\ V$

***Problems 22-2***

1. C is the load. $R_{TH} = R_3 + R_1 \| R_2 = 1\ k\Omega + 1.16\ k\Omega = 2.16\ k\Omega$

   $f_{co} = \dfrac{1}{2\pi R_{TH}C} = \dfrac{1}{2\pi \times 2.16\ k\Omega \times 500\ nF} = 147\ Hz$

   $V_{oc} = V_2 = \dfrac{ER_2}{R_1 + R_2} = \dfrac{10\ V \times 3.3\ k\Omega}{1.8\ k\Omega + 3.3\ k\Omega} = 6.47\ V$

$V_c = V_{max} \sin 45° = 4.57$ V, $I_c = I_3 = \dfrac{V_c}{X_c} = \dfrac{4.58 \text{ V}}{2.16 \text{ k}\Omega} = 2.11$ mA

$V_o = V_3 = I_3R_3 = 2.12 \text{ mA} \times 1 \text{ k}\Omega = 2.12$ V

3. $R_{TH} = R_2 = 2 \text{ k}\Omega$, $V_{max} = E = 20$ V

$f_{co} = \dfrac{1}{2\pi R_{TH}C} = \dfrac{1}{2\pi \times 2 \text{ k}\Omega \times 2 \ \mu\text{F}} = 39.8$ Hz, $V_o = V_{max} \sin 45° = 14.1$ V

5. $R_{TH} = R_1 + R_2 \| R_3 = 2.7 \text{ k}\Omega + 2.3 \text{ k}\Omega = 5.00 \text{ k}\Omega$, $V_{max} = E = 10$V

$f_{co} = \dfrac{1}{2\pi R_{TH}C} = \dfrac{1}{2\pi \times 5 \text{ k}\Omega \times 1 \ \mu\text{F}} = 31.8$ Hz

$V_{o(max)} = \dfrac{ER_x}{R_1 + R_x}$ where $R_x = R_2 \| R_3$, $V_{o(max)} = \dfrac{10 \text{ V} \times 2.3 \text{ k}\Omega}{2.3 \text{ k}\Omega + 2.7 \text{ k}\Omega} = 4.6$ V

$V_o$ at $f_{co} = V_{o(max)} \sin 45° = 3.25$ V

7. $R_{TH} = R_1 + R_2 \| R_3 = 2 \text{ k}\Omega + 2.08 \text{ k}\Omega = 4.08 \text{ k}\Omega$

$V_{max} = \dfrac{ER_3}{R_2 + R_3} = \dfrac{9 \times 5.6 \text{ k}\Omega}{8.9 \text{ k}\Omega} = 5.66$ V

$f_{co} = \dfrac{1}{2\pi R_{TH}C} = \dfrac{1}{2\pi \times 4.08 \text{ k}\Omega \times 1 \ \mu\text{F}} = 39$ Hz

9. $I_{SC} = \dfrac{E}{R_1} = \dfrac{15 \text{ V}}{470 \ \Omega} = 31.9$ mA

$G_N = G_1 + G_2 = 2.13 \text{ mS} + 833 \ \mu\text{S} = 2.96$ mS

$f_{co} = \dfrac{G_N}{2\pi C} = 9.43$ kHz, $V_{o(max)} = \dfrac{I_{SC}}{G_N} = 10.8$ V

$V_o$ at $f_{co} = 10.8 \text{ V} \times \sin 45° = 7.62$ V

11. $I_{SC} = \dfrac{E}{R_1} = \dfrac{10 \text{ V}}{2 \text{ k}\Omega} = 5$ mA, $G_N = G_1 + G_2 + G_3 = 891 \ \mu$S

$f_{co} = \dfrac{G_N}{2\pi C} = \dfrac{891 \ \mu\text{S}}{2\pi \times 1 \text{ nF}} = 142$ kHz

$V_{o(max)} = \dfrac{I_{SC}}{G_N} = \dfrac{5 \text{ mA}}{891 \ \mu\text{S}} = 5.61$ V, $V_o$ at $f_{co} = 5.61 \text{ V} \times \sin 45° = 3.97$ V

***Problems 22-3***

1. $R_{TH} = R_1 + R_2 = 1.88 \text{ k}\Omega$, $f_1 = \dfrac{1}{2\pi R_{TH}C_1} = \dfrac{1}{2\pi \times 1.88 \text{ k}\Omega \times 500 \text{ nF}} = 169$ Hz

$G_N = G_1 + G_2 = 1.47 \text{ mS} = 833 \ \mu\text{S} = 2.30$ mS

$f_2 = \dfrac{G_N}{2\pi C_2} = \dfrac{2.30 \text{ mS}}{2\pi \times 2 \text{ nF}} = 183$ kHz

$$V_o \text{ at } f_m = \frac{ER_2}{R_1 + R_2} = \frac{9\text{ V} \times 1.2\text{ k}\Omega}{680\ \Omega + 1.2\text{ k}\Omega} = 5.74\text{ V}$$

$$V_o \text{ at } f_{co} = V_{o(max)} \times \sin 45° = 4.06\text{ V}$$

3. $R_{TH} = R_1 + R_2 = 5.3\text{ k}\Omega$, $f_1 = \dfrac{1}{2\pi R_{TH} C_1} = \dfrac{1}{2\pi \times 5.3\text{ k}\Omega \times 2\ \mu\text{F}} = 15\text{ Hz}$

$G_N = G_1 + G_2 = 803\ \mu\text{S}$, $f_2 = \dfrac{G_N}{2\pi C_2} = \dfrac{803\ \mu\text{S}}{2\pi \times 1\text{ nF}} = 128\text{ kHz}$

$V_o$ at $f_m = \dfrac{ER_2}{R_1 + R_2} = 5.66\text{ V}$, $V_o$ at $f_{co} = 5.66\text{ V} \times \sin 45° = 4\text{ V}$

5. $R_{TH} = R_1 + R_2 = 6.1\text{ k}\Omega$, $f_1 = \dfrac{1}{2\pi R_{TH} C_1} = \dfrac{1}{2\pi \times 6.1\text{ k}\Omega \times 1.2\ \mu\text{F}} = 21.7\text{ Hz}$

$G_N = G_1 + G_2 = 711\ \mu\text{S}$, $f_2 = \dfrac{G_N}{2\pi C_2} = \dfrac{711\ \mu\text{S}}{2\pi \times 20\text{ nF}} = 5.66\text{ kHz}$

$V_o$ at $f_m = \dfrac{ER_2}{R_1 + R_2} = 1.28\text{ V}$, $V_o$ at $f_{co} = 1.28\text{ V} \times \sin 45° = 904\text{ mV}$

7. $R_{TH} = R_1 + R_2 = 1.15\text{ k}\Omega$, $f_1 = \dfrac{1}{2\pi R_{TH} C_1} = \dfrac{1}{2\pi \times 1.15\text{ k}\Omega \times 200\text{ nF}} = 692\text{ Hz}$

$G_N = G_1 + G_2 = 4.25\text{ mS}$, $f_2 = \dfrac{G_N}{2\pi C_2} = \dfrac{4.25\text{ mS}}{2\pi \times 500\text{ pF}} = 1.35\text{ MHz}$

$V_o$ at $f_m = \dfrac{ER_2}{R_1 + R_2} = 2.14\text{ V}$, $V_o = 2.14\text{ V} \times \sin 45° = 1.51\text{ V}$

9. $R_{TH} = R_3 + R_1 \| R_2 = 1.8\text{ k}\Omega + 1.82\text{ k}\Omega = 3.62\text{ k}\Omega$

$f_1 = \dfrac{1}{2\pi R_{TH} C_1} = \dfrac{1}{2\pi \times 3.62\text{ k}\Omega \times 1\ \mu\text{F}} = 43.9\text{ Hz}$

$G_N = G_1 + G_2 + G_3 = 1.10\text{ mS}$, $F_2 = \dfrac{G_N}{2\pi C_2} = \dfrac{1.10\text{ mS}}{2\pi \times 2\text{ nF}} = 87.9\text{ kHz}$

11. $R_{TH} = R_3 + R_1 \| R_2 = 6.8\text{ k}\Omega + 938\ \Omega = 7.74\text{ k}\Omega$

$f_1 = \dfrac{1}{2\pi R_{TH} C_1} = \dfrac{1}{2\pi \times 7.74\text{ k}\Omega \times 2\ \mu\text{F}} = 10.3\text{ Hz}$

$G_N = G_1 + G_2 + G_3 = 1.21\text{ mS}$, $f_2 = \dfrac{G_N}{2\pi C_2} = \dfrac{1.21\text{ mS}}{2\pi \times 800\text{ pF}} = 241\text{ kHz}$

13. $BW = \dfrac{f_r}{Q} = \dfrac{2\text{ kHz}}{25} = 80\text{ Hz}$

$f_1 = f_r - \dfrac{BW}{2} = 2\text{ kHz} - 40\text{ Hz} = 1.96\text{ kHz}$

$f_2 = f_r + \frac{BW}{2} = 2\ kHz + 40\ Hz = 2.04\ kHz$

at $f_r$: $V_R = V_{in} = 20\ V$, $I = \frac{V_{in}}{R} = \frac{20\ V}{1\ k\Omega} = 20\ mA$; $P = V_{in} \times I = 400\ mW$

at $f_1$ and $f_2$: $V_R = V_{R(max)} \sin 45° = 20\ V \times \sin 45° = 14.1\ V$

$$I = \frac{V_R}{R} = \frac{14.1\ V}{1\ k\Omega} = 14.1\ mA$$

$$P = IV_R = 14.1\ mA \times 14.1\ V = 200\ mW$$

15. $BW = \frac{f_r}{Q} = \frac{5.5\ kHz}{14} = 393\ Hz$

$f_1 = f_r - \frac{BW}{2} = 5.30\ kHz$, $f_2 = f_r + \frac{BW}{2} = 5.70\ kHz$

at $f_r$: $V_R = V_{in} = 5\ V$, $I = \frac{V_{in}}{R} = \frac{5\ V}{120\ \Omega} = 41.7\ mA$;

$$P = V_{in}\ I = 5\ V \times 41.7\ mA = 208\ mW$$

at $f_1$ and $f_2$: $V_R = V_{R(max)} \sin 45° = 5\ V \times \sin 45° = 3.54\ V$

$$I = \frac{V_R}{R} = \frac{3.54\ V}{120\ \Omega} = 29.5\ mA$$

$$P = V_R\ I = 3.54\ V \times 29.5\ mA = 104\ mW$$

17. $f_r = \frac{1}{2\pi(LC)^{1/2}} = \frac{1}{2\pi(100\ mH \times 250\ nF)^{1/2}} = 1.01\ kHz$

$X_L = 2\pi fL = 2\pi \times 1.01\ kHz \times 100\ mH = 632\ \Omega$

$Q = \frac{X_L}{R} = \frac{632\ \Omega}{45\ \Omega} = 14.1$

$BW = \frac{f_r}{Q} = \frac{1.01\ kHz}{14.1} = 71.9\ Hz$

$f_1 = f_r - \frac{BW}{2} = 974\ Hz.$ $\quad f_2 = f_2 + \frac{BW}{} = 1.05\ kHz$

19. $f_r = \frac{1}{2\pi(LC)^{1/2}} = 3.56\ kHz$, $X_L = 2\pi fL = 447\ \Omega$

$Q = \frac{X_L}{R} = \frac{447\ \Omega}{75\ \Omega} = 5.96$, $BW = \frac{f_r}{Q} = 597\ Hz$

$f_1 = f_r - \frac{BW}{2} = 3.56\ kHz - 299\ Hz = 3.26\ kHz$

$f_2 = f_r - \frac{BW}{2} = 3.56\ kHz + 299\ Hz = 3.86\ kHz$

# CHAPTER 23

***Problems 23-1***

| | *Logarithmic Form* | *Exponential Form* |
|---|---|---|
| 1. | log 1000 = 3 | $10^{3}$ = 1,000 |
| 3. | log 1 = 0 | $10^{0}$ = 1 |
| 5. | log 10,000,000 = 7 | $10^{7}$ = 10,000,000 |
| 7. | log 0.1 = -1 | $10^{-1}$ = 0.1 |
| 9. | log 0.000001 = -6 | $10^{-6}$ = 0.000001 |

***Problems 23-2***

| | *Logarithmic Form* | *Exponential Form* |
|---|---|---|
| 1. | log 3.42 = 0.534 | $10^{0.534}$ = 3.42 |
| 3. | log 4.05 = 0.608 | $10^{0.608}$ = 4.05 |
| 5. | log 6.44 = 0.809 | $10^{0.809}$ = 6.44 |
| 7. | log 9.45 = 0.975 | $10^{0.975}$ = 9.45 |
| 9. | log 2.50 = 0.398 | $10^{0.398}$ = 2.50 |

***Problems 23-3***

| | *Logarithmic Form* | *Exponential Form* |
|---|---|---|
| 1. | log 14 = 1.15 | $10^{1.15}$ = 14 |
| 3. | log 743 = 2.87 | $10^{2.87}$ = 743 |
| 5. | log 1,070 = 3.03 | $10^{3.03}$ = 1,070 |
| 7. | log 43.4 = 1.64 | $10^{1.64}$ = 43.4 |
| 9. | log 10,000 = 4.00 | $10^{4.00}$ = 10,000 |
| 11. | log 23,400 = 4.37 | $10^{4.37}$ = 23,400 |
| 13. | log 47,700 = 4.68 | $10^{4.68}$ = 47,700 |
| 15. | log 14.6 = 1.16 | $10^{1.16}$ = 14.6 |
| 17. | log 37,000 = 4.57 | $10^{4.57}$ = 37,000 |
| 19. | log 2,000,000 = 6.30 | $10^{6.30}$ = 2,000,000 |

***Problems 23-4***

| | *Logarithmic Form* | *Exponential Form* |
|---|---|---|
| 1. | log 0.00377 = -2.42 | $10^{-2.42}$ = 0.00377 |
| 3. | log 0.037 = -1.43 | $10^{-1.43}$ = 0.037 |
| 5. | log 0.746 = -0.127 | $10^{-0.127}$ = 0.746 |
| 7. | log 0.407 = -0.390 | $10^{-0.390}$ = 0.407 |
| 9. | log 0.000614 = -3.21 | $10^{-3.21}$ = 0.000614 |
| 11. | log 0.49 = -0.310 | $10^{-0.310}$ = 0.49 |
| 13. | log 0.00814 = -2.09 | $10^{-2.09}$ = 0.00814 |
| 15. | log 0.301 = -0.521 | $10^{-0.521}$ = 0.301 |
| 17. | log 0.975 = -0.011 | $10^{-0.011}$ = 0.975 |
| 19. | log 0.0275 = -1.56 | $10^{-1.56}$ = 0.0275 |

***Problems 23-5***

1. log 27 + log 56
   = 1.431 + 1.748 = 3.179
   $10^{3.179} = 1.510 \times 10^{3}$
5. log 43 + log 0.143
   = 1.633 + (-0.8447) = 0.7888
   $10^{0.7888} = 6.15$
9. log 0.0000176 + log 0.000357
   = -4.754 + (-3.447) = -8.202
   $10^{-8.202} = 6.28 \times 10^{-9}$
13. log 78 + log 0.473
   = 1.892 + (-0.3251) = 1.567
   $10^{1.567} = 36.9$
17. log 760 + log 7360
   = 2.881 + 3.867 = 6.748
   $10^{6.748} = 5.59 \times 10^{6}$

3. log 107 + log 243,000
   = 2.029 + 5.386 = 7.415
   $10^{7.415} = 2.60 \times 10^{7}$
7. log 0.043 + log 0.0106
   = -1.367 + (-1.975) = -3.341
   $10^{-3.341} = 4.56 \times 10^{-4}$
11. log 256 + log 0.00246
   = 2.408 + (-2.609) = -0.2008
   $10^{-0.2008} = 6.30 \times 10^{-1}$
15. log 4730 + log 0.000506
   = 3.675 + -3.296 = 0.3790
   $10^{0.3790} = 2.39$
19. log 2560 + log 0.00457
   = 3.408 + (-2.340) = 1.068
   $10^{1.068} = 11.7$

***Problems 23-6***

1. log 1760 - log 43
   = 3.246 - 1.633 = 1.612
   $10^{1.613} = 40.9$
5. log 0.00273 - log 172
   = -2.564 - 2.236 = -4.799
   $10^{-4.80} = 1.59 \times 10^{-5}$
9. log 0.0675 - log 0.00103
   = -1.171 - (-2.987) = 1.816
   $10^{1.816} = 65.5$
13. log 0.147 - log 0.0432
   = -0.8327 - (-1.365) = 0.5318
   $10^{0.5318} = 3.40$
17. log 0.445 - log 635
   = -0.3516 - 2.803 = -3.154
   $10^{-3.154} = 7.01 \times 10^{-4}$

3. log 65 - log 467
   = 1.813 - 2.669 = -0.8563
   $10^{-0.8563} = 1.39 \times 10^{-1}$
7. log 4300 - log 0.15
   = 3.633 - (-0.824) = 4.457
   $10^{4.457} = 2.87 \times 10^{4}$
11. log 0.00467 - log 73.2
   = -2.331 - 1.865 = -4.196
   $10^{-4.196} = 6.38 \times 10^{-5}$
15. log 4730 - log 0.443
   = 3.675 - (-0.3536) = 4.029
   $10^{4.029} = 1.07 \times 10^{4}$
19. log 0.00275 - log 4.35
   = -2.561 - 0.6385 = -3.199
   $10^{-3.199} = 6.32 \times 10^{-4}$

***Problems 23-7***

1. 3 log 3.4 = 3 x 0.5315
   = 1.594
   $10^{1.594} = 39.3$
5. 2.5 log 0.43 = 2.5 x -0.3665
   = -0.9163
   $10^{-0.9163} = 1.21 \times 10^{-1}$

3. -1.3 log 6 = -1.3 x 0.7782
   = -1.012
   $10^{-1.012} \times 9.74 \times 10^{-2}$
7. 4.3 log 6 = 4.3 x 0.7782
   = 3.346
   $10^{3.346} = 2.22 \times 10^{3}$

9. $-4.4 \log 0.27 = -4.4 \times -0.5686$
$= 2.502$
$10^{2.502} = 3.18 \times 10^{2}$

11. $6.3 \log 8.7 = 6.3 \times 0.9395$
$= 5.919$
$10^{5.919} = 8.30 \times 10^{5}$

13. $3.2 \log 3.4 = 3.2 \times 0.5315$
$= 1.701$
$10^{1.701} = 50.2$

15. $-2.3 \log 6.35 = -2.3 \times 0.8028$
$= -1.846$
$10^{-1.846} = 1.42 \times 10^{-2}$

17. $3.6 \log 0.045 = 3.6 \times -1.347$
$= -4.848$
$10^{-4.848} = 1.42 \times 10^{-5}$

19. $-2.5 \log 0.00735 = -2.5 \times -2.134$
$= 5.334$
$10^{5.334} = 2.16 \times 10^{5}$

***Problems 23-8***

| | *Logarithmic Form* | *Exponential Form* |
|---|---|---|
| 1. | $\ln 2 = 0.6931$ | $e^{0.6931} = 2$ |
| 3. | $\ln 0.932 = -0.07042$ | $e^{-0.07042} = 0.932$ |
| 5. | $\ln 60 = 4.094$ | $e^{4.094} = 60$ |
| 7. | $\ln 0.073 = -2.617$ | $e^{-2.617} = 0.073$ |
| 9. | $\ln 247 = 5.509$ | $e^{5.509} = 247$ |
| 11. | $\ln 0.00146 = -6.529$ | $e^{-6.529} = 0.00146$ |
| 13. | $\ln 1100 = 7.003$ | $e^{7.003} = 1100$ |
| 15. | $\ln 0.00027 = -8.217$ | $e^{-8.217} = 0.00027$ |
| 17. | $\ln 8.43 = 2.132$ | $e^{2.132} = 8.43$ |
| 19. | $\ln 0.346 = -1.061$ | $e^{-1.061} = 0.346$ |
| 21. | $\ln 3.67 = 1.30$ | $e^{1.30} = 3.67$ |
| 23. | $\ln 0.107 = -2.235$ | $e^{-2.235} = 0.107$ |
| 25. | $\ln 176 = 5.170$ | $e^{5.170} = 176$ |
| 27. | $\ln 0.293 = -1.228$ | $e^{-1.228} = 0.293$ |
| 29. | $\ln 873 = 6.772$ | $e^{6.772} = 873$ |

***Problems 23-9***

1. $e^{2.34} = 10.38$
3. $e^{3.77} = 43.38$
5. $e^{4.25} = 70.11$
7. $e^{5.3} = 200.3$
9. $e^{1.56} = 4.711$
11. $e^{0.43} = 1.537$
13. $e^{-0.0045} = 0.9955$
15. $e^{0.15} = 1.162$
17. $e^{-0.0315} = 0.7298$
19. $e^{-0.142} = 0.8676$

# CHAPTER 24

***Problems 24-1***

1. $10^x = 340$
3. $10^x = 14$
5. $10^x = 7.4$
7. $10^y = 0.014$
9. $10^y = 0.734$
11. $10^y = -0.00178$
13. $x = 10^{-2.43}$
15. $x = 10^{-3.643}$
17. $x = 10^{-0.143}$
19. $y = 10^{1.44}$
21. $y = 10^{2.804}$
23. $y = 10^{4.146}$
25. $\log y = 3.6$
27. $\log y = 4.73$
29. $\log y = 2.73$
31. $a = \log 732$
33. $a = \log 47.3$
35. $a = \log 276$
37. $y = \log 0.043$
39. $y = \log 0.1778$
41. $y = \log 0.00178$
43. $\log x = -3.4$
45. $\log x = -2.76$
47. $\log x = -1.17$

***Problems 24-2***

1. $y = 2.291 \times 10^7$
3. $x = 1.002$
5. $a = 0.3776$
7. $x = 0.9917$
9. $x = 2.665$
11. $x = -1.762$
13. $y = -2.602$
15. $y = 4.668$
17. $x = 1.445 \times 10^4$
19. $x = 1.079$
21. $y = 2.512 \times 10^{-4}$
23. $y = 0.6714$
25. $x = 2.873$
27. $x = -2.145$
29. $x = 2.778 \times 10^3$
31. $x \log 0.026 = \log 7$

    $x = \dfrac{\log 7}{\log 0.026} = -0.5332$

33. $4.3 \log x = \log 7300$

    $\log x = \dfrac{\log 7300}{4.3} = 0.8984$

    $10^{0.8984} = x = 7.915$

35. $3 \log y = 9$

    $\log y = \dfrac{9}{3} = 3$

    $y = 10^3 = 1 \times 10^3$

37. $10^{7.36} = y = 2.291 \times 10^7$
39. $7 \log x = \log 14$

    $\log x = \dfrac{\log 14}{7} = 0.1637$

    $10^{0.1637} = x = 1.458$

41. $x \log 70 = \log 12$

    $x = \dfrac{\log 12}{\log 70} = 0.5849$

***Problems 24-3***

1. $(2 - y) \log 6 = \log 68$

    $2 - y = \dfrac{\log 68}{\log 6} = 2.355$

    $2 - 2.355 = y$, $y = -0.3550$

3. $\dfrac{y}{4} \log 75 = \log 3.54$

    $\dfrac{y}{4} = \dfrac{\log 3.54}{\log 75} = 0.2928$

    $y = 1.171$

5. $10^{1.7} = \frac{x}{42}$

$x = 42 \times 10^{1.7} = 2.105 \times 10^{3}$

7. $10^{6} = \frac{100}{x}$

$x = \frac{100}{10^{6}} = 1 \times 10^{-4}$

9. $(x - 2) \log 3 = \log 50$

$x - 2 = \frac{\log 50}{\log 3} = 3.561$

$x = 5.561$

11. $\frac{3}{x} \log 16 = \log 200$

$\frac{3}{x} = \frac{\log 200}{\log 16} = 1.911$

$\frac{3}{1.911} = x, \ x = 1.570$

13. $10^{6} = \frac{x}{100}$

$x = 100 \times 10^{6} = 1 \times 10^{8}$

15. $10^{-4} = \frac{20}{x}$

$x = \frac{20}{10^{-4}} = 2 \times 10^{5}$

17. $\frac{x}{2} \log 14 = \log 43$

$\frac{x}{2} = \frac{\log 43}{\log 14} = 1.425$

$x = 2.850$

19. $(3 - x) \log 35 = \log 470$

$3 - x = \frac{\log 470}{\log 35} = 1.731$

$3 - 1.731 = x, \ x = 1.269$

21. $10^{3} = \frac{0.005}{x}$

$x = \frac{0.005}{10^{3}} = 5 \times 10^{-6}$

23. $10^{4.3} = \frac{y}{9}$

$9 \times 10^{4.3} = y$

$y = 1.796 \times 10^{5}$

25. $(x - 2) \log 45 = \log 3.75$

$x - 2 = \frac{\log 3.75}{\log 45} = 0.3472$

$x = 2.347$

27. $(x + 2) \log 8.7 = \log 150$

$x + 2 = \frac{\log 150}{\log 8.7}$

$x = 0.3162$

29. $\frac{y}{3} \log 60 = \log 1.78$

$\frac{y}{3} = \frac{\log 1.78}{\log 60}$

$\frac{y}{3} = 0.1408, \ y = 0.4225$

*Problems 24-4*

1. $x = 2.398$

3. $x = 2.708$

5. $x = 5.298$

7. $x = -7.824$

9. $x = -2.708$

11. $x = 14.88$

13. $y = 3.829$

15. $y = 1.832 \times 10^{-2}$

***Problems 24-5***

1. $0.6254 - 1 = -e^{-x}$

   $e^{-x} = 0.3746$

   $-x \ln e = \ln 0.3746$

   $-x = -0.9819,\ x = 0.9819$

3. $\frac{50}{80} = e^{-x}$

   $e^{-x} = 0.625$

   $-x \ln e = \ln 0.625$

   $-x = -0.47,\ x = 0.47$

5. $x = 25(1 - 0.4724)$

   $x = 13.19$

7. $30 = x(1 - 4.979 \times 10^{-2})$

   $30 = x(0.9502)$

   $x = \frac{30}{0.9502} = 31.57$

9. $\frac{10}{25} = e^{-x/3}$

   $0.4 = e^{-x/3}$

   $\ln 0.4 = -\frac{x}{3} \ln e$

   $-0.9163 = -\frac{x}{3}$

   $\frac{x}{3} = 0.9163$

   $x = 2.749$

11. $\frac{3}{9} = e^{-4/x}$

    $0.333 = e^{-4/x}$

    $\ln 0.333 = -\frac{4}{x} \ln e$

    $-1.099 = -\frac{4}{x}$

    $x = \frac{-4}{-1.099} = 3.641$

13. $\ln 0.176 = \frac{-2x}{3} \ln e$

    $-1.737 = \frac{-2x}{3}$

    $3(1.737) = 2x$

    $x = \frac{5.211}{2} = 2.606$

15. $\ln 0.5 = \frac{-2}{x} \ln e$

    $-0.6931 = \frac{-2}{x}$

    $x = \frac{2}{0.6931} = 2.885$

17. $x = 1 - 4.979 \times 10^{-2} = 0.9502$

19. $0.376 - 1 = -e^{-x}$

    $0.6240 = e^{-x}$

    $\ln 0.624 = -x \ln e$

    $-0.4716 = -x$

    $x = 0.4716$

21. $\frac{-x}{2} \ln e = \ln 0.76$

    $\frac{-x}{2} = -0.2744$

    $x = 0.5489$

# CHAPTER 25

***Problems 25-1***

1. $A_p = 10 \log \dfrac{P_o}{P_{in}} = 60 \text{ dB}$

3. $A_p = 10 \log \dfrac{P_o}{P_{in}} = 43 \text{ dB}$

5. $A_p = 10 \log \dfrac{P_o}{P_{in}} = 44 \text{ dB}$

7. $P_{in} = \dfrac{P_o}{10^{3.5}} = 3.79 \text{ mW}$

9. $P_{in} = \dfrac{P_o}{10^{2}} = 1 \text{ W}$

11. $P_{in} = \dfrac{P_o}{10^{3}} = 5 \text{ mW}$

13. $P_o = P_{in} \times 10^{2.2} = 1.58 \text{ W}$

15. $P_o = P_{in} \times 10^{2.8} = 3.15 \text{ W}$

17. $P_o = P_{in} \times 10^{1.5} = 316 \text{ mW}$

19. $P_o = 1 \text{ mW} \times 10^{2} = 100 \text{ mW}$

21. $P_o = 1 \text{ mW} \times 10^{3.2} = 1.58 \text{ W}$

23. $P_o = 1 \text{ mW} \times 10^{5} = 100 \text{ W}$

25. $A_v = 20 \log \dfrac{2.7 \text{ V}}{100 \ \mu\text{V}} = 88.6 \text{ dB}$

27. $A_v = 20 \log \dfrac{3 \text{ V}}{10 \text{ mV}} = 49.5 \text{ dB}$

29. $A_v = 20 \log \dfrac{10 \text{ V}}{50 \ \mu\text{V}} = 106 \text{ dB}$

31. $V_o = 5 \text{ mV} \times 10^{2} = 500 \text{ mV}$

33. $V_o = 2 \text{ mV} \times 10^{1.14} = 27.3 \text{ mV}$

35. $V_o = 23 \text{ mV} \times 10^{3.65} = 103 \text{ V}$

37. $V_{in} = \dfrac{4 \text{ V}}{10^{4.5}} = 126 \ \mu\text{V}$

39. $V_{in} = \dfrac{15 \text{ V}}{10^{3.75}} = 2.67 \text{ mV}$

41. $V_{in} = \dfrac{600 \text{ mV}}{10^{4}} = 60 \ \mu\text{V}$

43. $V_{in} = \dfrac{500 \text{ mV}}{10^{-0.15}} = 706 \text{ mV}$

45. $V_{in} = \dfrac{1 \text{ V}}{10^{-0.3}} = 2 \text{ V}$

47. $V_{in} = \frac{5\text{ V}}{10} = 7.06\text{ V}$

49. $V_{in} = \frac{100\text{ mV}}{10} = 3.16\text{ V}$

***Problems 25-2***

1. $f_{co} = \frac{1}{2\pi RC} = \frac{1}{2\pi \times 1\text{ k}\Omega \times 0.3\ \mu\text{F}} = 531\text{ Hz}$

3. $f_{co} = \frac{1}{2\pi RC} = \frac{1}{2\pi \times 470\ \Omega \times 1\ \mu\text{F}} = 339\text{ Hz}$

***Problems 25-2***

1 & 3.

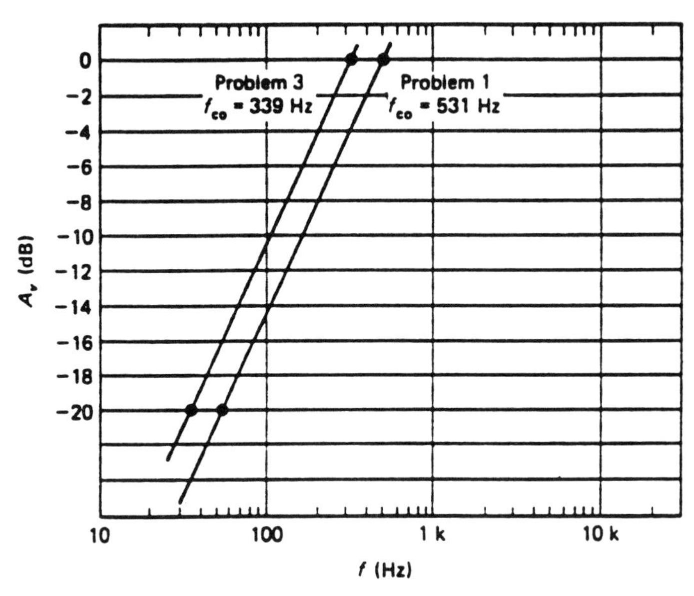


5. $f_{co} = \frac{1}{2\pi RC} = \frac{1}{2\pi \times 3.9\text{ k}\Omega \times 300\text{ nF}} = 136\text{ Hz}$

7. $f_{co} = \frac{1}{2\pi RC} = \frac{1}{2\pi \times 2\text{ k}\Omega \times 50\text{ nF}} = 1.59\text{ kHz}$

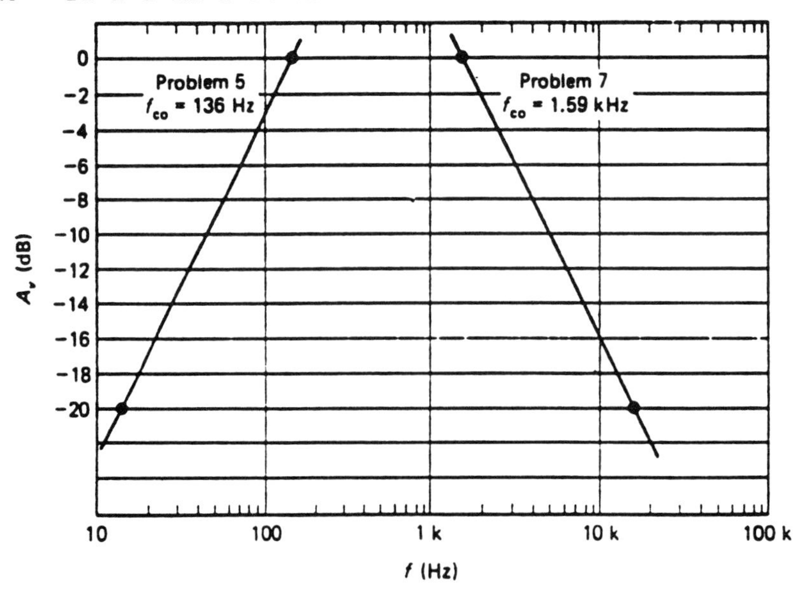

***Problems 25-2***

9. $f_{co} = \frac{1}{2\pi RC} = \frac{1}{2\pi \times 750\ \Omega \times 200\ nF} = 1.06\ kHz$

11. $f_{co} = \frac{1}{2\pi RC} = \frac{1}{2\pi \times 2\ k\Omega \times 40\ nF} = 1.99\ kHz$

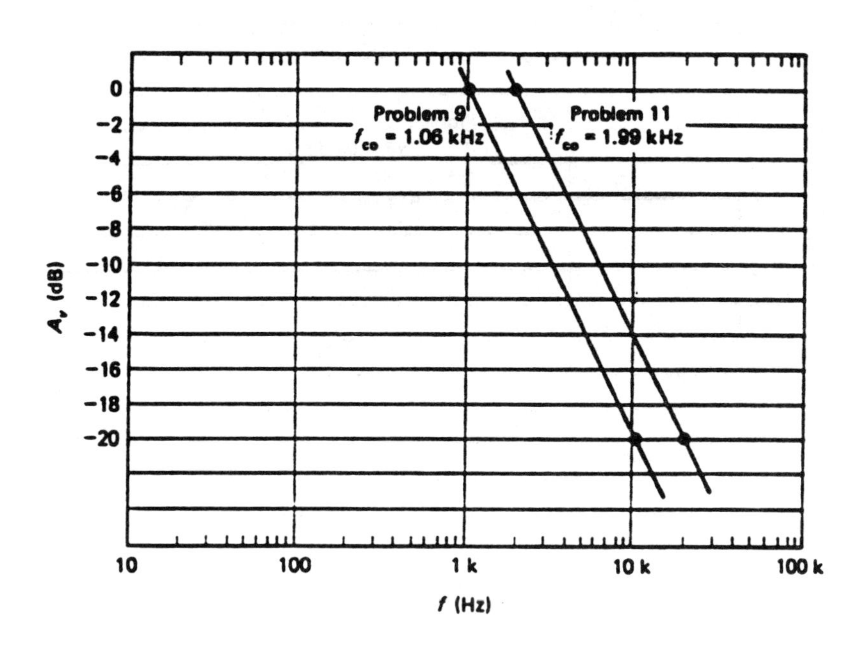


***Problems 25-3***

1. $f_1 = \frac{1}{2\pi(R_1 + R_2)C_1} = \frac{1}{2\pi \times 1.88\ k\Omega \times 250\ nF} = 339\ Hz$

$f_2 = \frac{1}{2\pi(R_1 \| R_2)C_2} = \frac{1}{2\pi \times 434\ \Omega \times 40\ nF} = 9.17\ kHz$

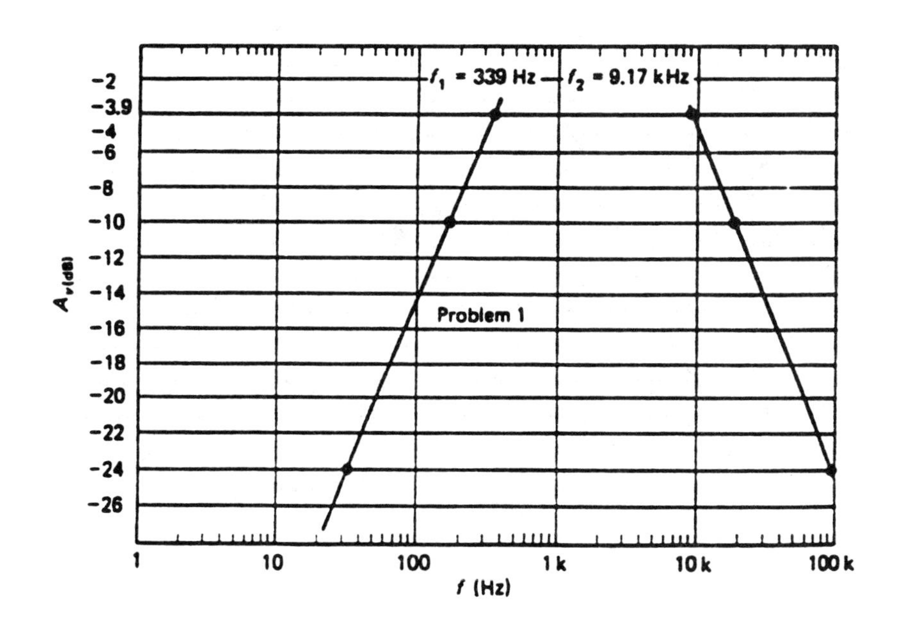

***Problems 25-3***

3. $$f_1 = \frac{1}{2\pi(R_1 + R_2)C_1} = \frac{1}{2\pi \times 2.27\ \text{k}\Omega \times 500\ \text{nF}} = 140\ \text{Hz}$$

$$f_2 = \frac{1}{2\pi(R_1 \| R_2)C_2} = \frac{1}{2\pi \times 373\ \Omega \times 50\ \text{nF}} = 8.54\ \text{kHz}$$

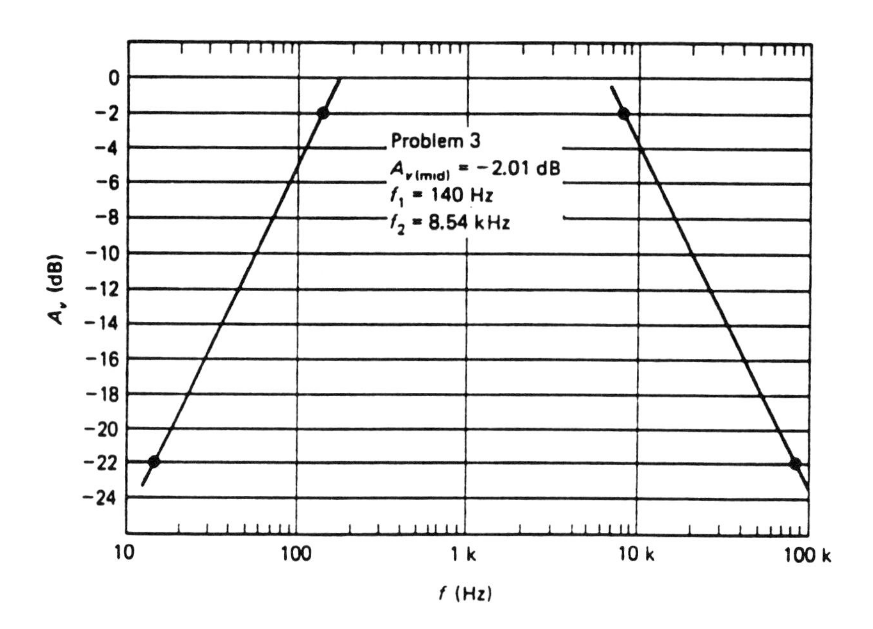


5. $$f_1 = \frac{1}{2\pi(R_1 + R_2)C_1} = \frac{1}{2\pi \times 5.5\ \text{k}\Omega \times 200\ \text{nF}} = 145\ \text{Hz}$$

$$f_2 = \frac{1}{2\pi(R_1 \| R_2)C_2} = \frac{1}{2\pi \times 1.32\ \text{k}\Omega \times 30\ \text{nF}} = 4.02\ \text{kHz}$$

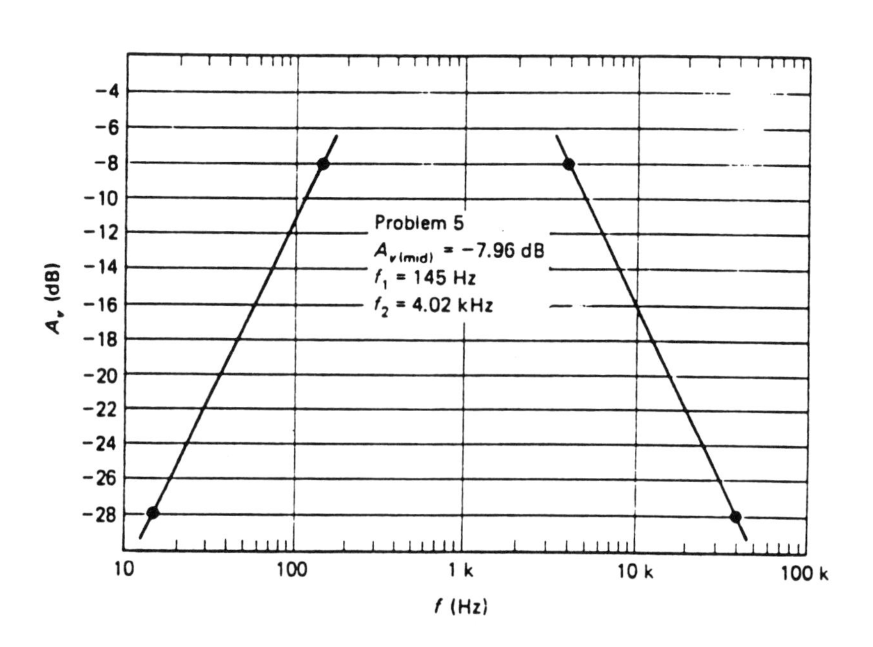

***Problems 25-3***

7. At 40 Hz (gen. case): $A_v = 20 \log \frac{1}{[1 + (f_1/f)^2]^{1/2}} = 20 \log \frac{1}{1.60} = -4.09$ dB

$A_v$ = 20 dB - 4.09 dB = 15.9 dB

At 25 kHz (gen. case): $A_v = 20 \log \frac{1}{[1 + (f/f_2)^2]^{1/2}} = 20 \log \frac{1}{1.71} = -4.67$ dB

$A_v$ = 20 dB - 4.67 dB = 15.3 dB

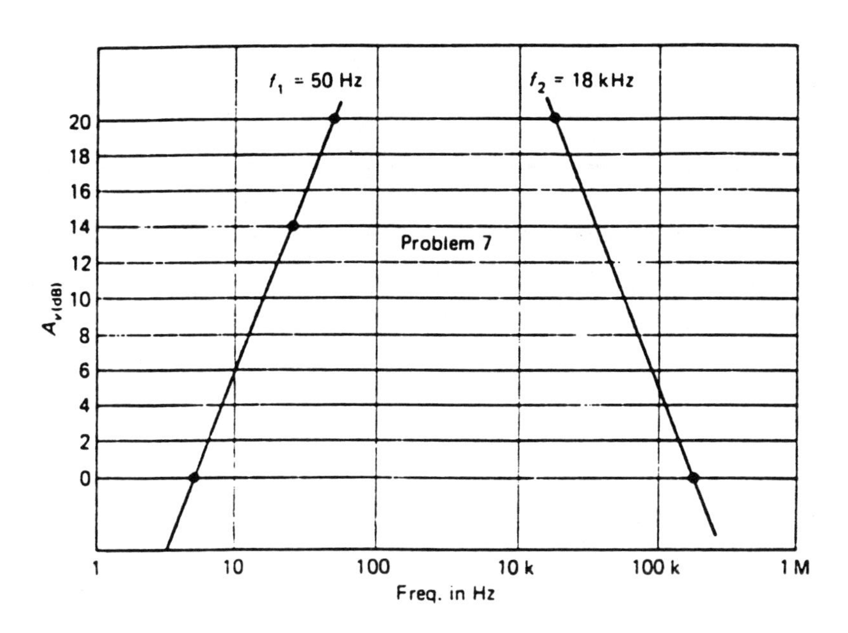


9. At 40 Hz (gen. case): $A_v = 20 \log \frac{1}{[1 + (f_1/f)^2]^{1/2}} = 20 \log \frac{1}{2.24} = -6.99$ dB

$A_v$ = 10 dB - 6.99 dB = 3.01 dB

At 12 kHz (gen. case): $A_v = 20 \log \frac{1}{[1 + (f/f_2)^2]^{1/2}} = 20 \log \frac{1}{3.16} = -10$ dB

$A_v$ = 10 dB - 10 dB = 0 dB

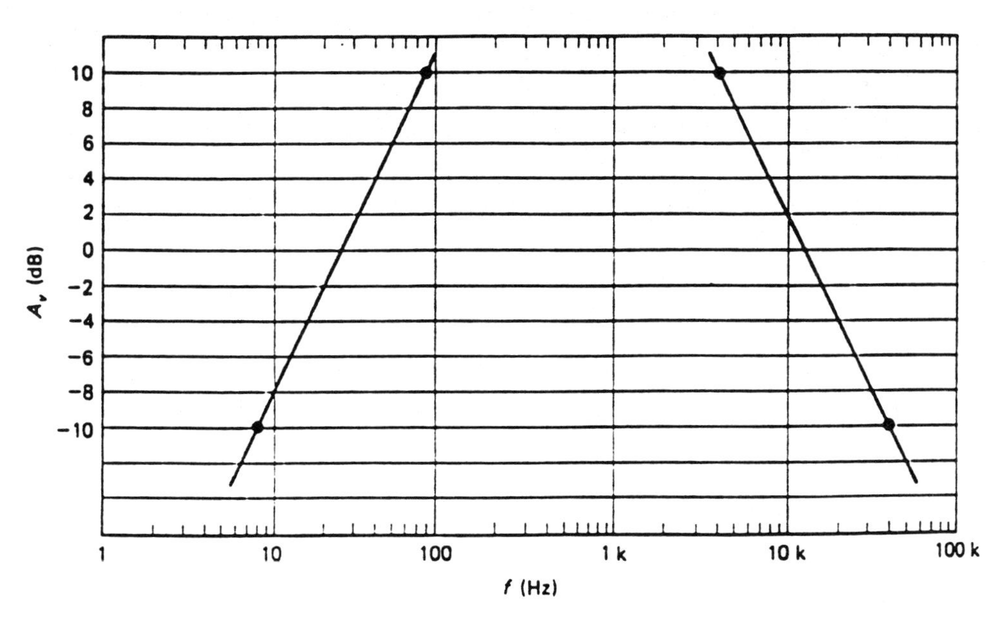

***Problems 25-3***

11. $f_1 = \dfrac{1}{2\pi(R_1 + R_2)C_1} = \dfrac{1}{2\pi \times 1.15\ k\Omega \times 1\ \mu F} = 138\ Hz$

$f_2 = \dfrac{1}{2\pi(R_1 \| R_2)C_2} = \dfrac{1}{2\pi \times 278\ \Omega \times 20\ nF} = 28.6\ kHz$

$A_{VM} = 20 \log \dfrac{R_2}{R_1 + R_2} = 20 \log \dfrac{680}{1.15\ k\Omega} = -4.56\ dB$

At 30 Hz: $A_v = -4.65\ dB + 20 \log \dfrac{1}{[1 + (f_1/f)^2]^{1/2}} = -4.65\ dB + 20 \log \dfrac{1}{4.71}$

$= -4.56\ dB - 13.5\ dB = -18.1\ dB$

At 50 kHz: $A_v = -4.56\ dB + 20 \log \dfrac{1}{[1 + (f/f_2)^2]^{1/2}} = -4.56\ dB + 20 \log \dfrac{1}{2.0}$

$= -4.56\ dB - 6.08\ dB = -10.6\ dB$

$TC = RC = 47\ k\Omega \times 50\ nF = 2.35\ ms.$ $I_{MAX} = \dfrac{100\ V}{47\ k\Omega} = 2.13\ mA$

***Problems 25-4***

1. $v_R = Ee^{-t/RC} = 100\ V \times e^{-1\ ms/2.35\ ms} = 65.3\ V$

$v_C = E(1 - e^{-t/RC}) = 100\ V(1 - e^{-1\ ms/2.35\ ms}) = 34.7\ V$

3. $65\ V = 100\ V(1 - e^{-t/2.35\ ms})$

$\dfrac{65\ V}{100\ V} = 1 - e^{-t/2.35\ ms}$

$-0.350 = -\ e^{-t/2.35\ ms}$

$0.350 = e^{-t/2.35\ ms}$

$\ln 0.35 = -t/2.35\ ms \times \ln e$

$-1.05 = -t/2.35\ ms$

$2.35\ ms \times 1.05 = t$

$t = 2.47\ ms$

5. $2\ mA = 2.13\ mA \times e^{-t/2.35\ ms}$

$0.939 = -\ e^{-t/2.35\ ms}$

$\ln 0.939 = -t/2.35\ ms \times \ln e$

$-61.9 \times 10^{-3} = -t/2.35\ ms$

$2.35\ ms \times 61.9 \times 10^{-3} = t$

$t = 145\ \mu s$

7. $10\ V = 100\ V \times e^{-t/2.35\ ms}$

$0.100 = -\ e^{-t/2.35\ ms}$

$\ln 0.100 = -t/2.35\ ms \times \ln e$

$-2.3 = -t/2.35\ ms$

$2.35\ ms \times 2.3 = t$

$t = 5.41\ ms$

9. $v_R = Ee^{-t/RC}$

$15\ V = 25\ V \times e^{-50\ \mu s/(10\ nF \times R)}$

$0.6 = e^{-50\ \mu s/(10\ nF \times R)}$

$\ln 0.6 = -50\ \mu s/(10\ nF \times R) \times \ln e$

$$R = \frac{50\ \mu s}{511 \times 10^{-3} \times 10\ nF} = 9.76\ k\Omega$$

11. $20\ V = 100\ V(1 - e^{-100\ \mu s/(47\ k\Omega \times C)}$

$0.2 = 1 - e^{-100\ \mu s/(47\ k\Omega \times C)}$

$-0.800 = -\ e^{-100\ \mu s/47\ k\Omega \times C)}$

$0.800 = e^{-100/(47\ k\Omega \times C)}$

$\ln 0.8 = -100 \mu s/(47\ k\Omega \times C) \times \ln e$

$-223 \times 10^{-3} = 100\ \mu s/(47\ k\Omega \times C)$

$$C = \frac{-100\ \mu s}{-223 \times 10^{-3} \times 47\ k\Omega} = 9.53\ nF$$

# CHAPTER 26

***Problems 26-1***

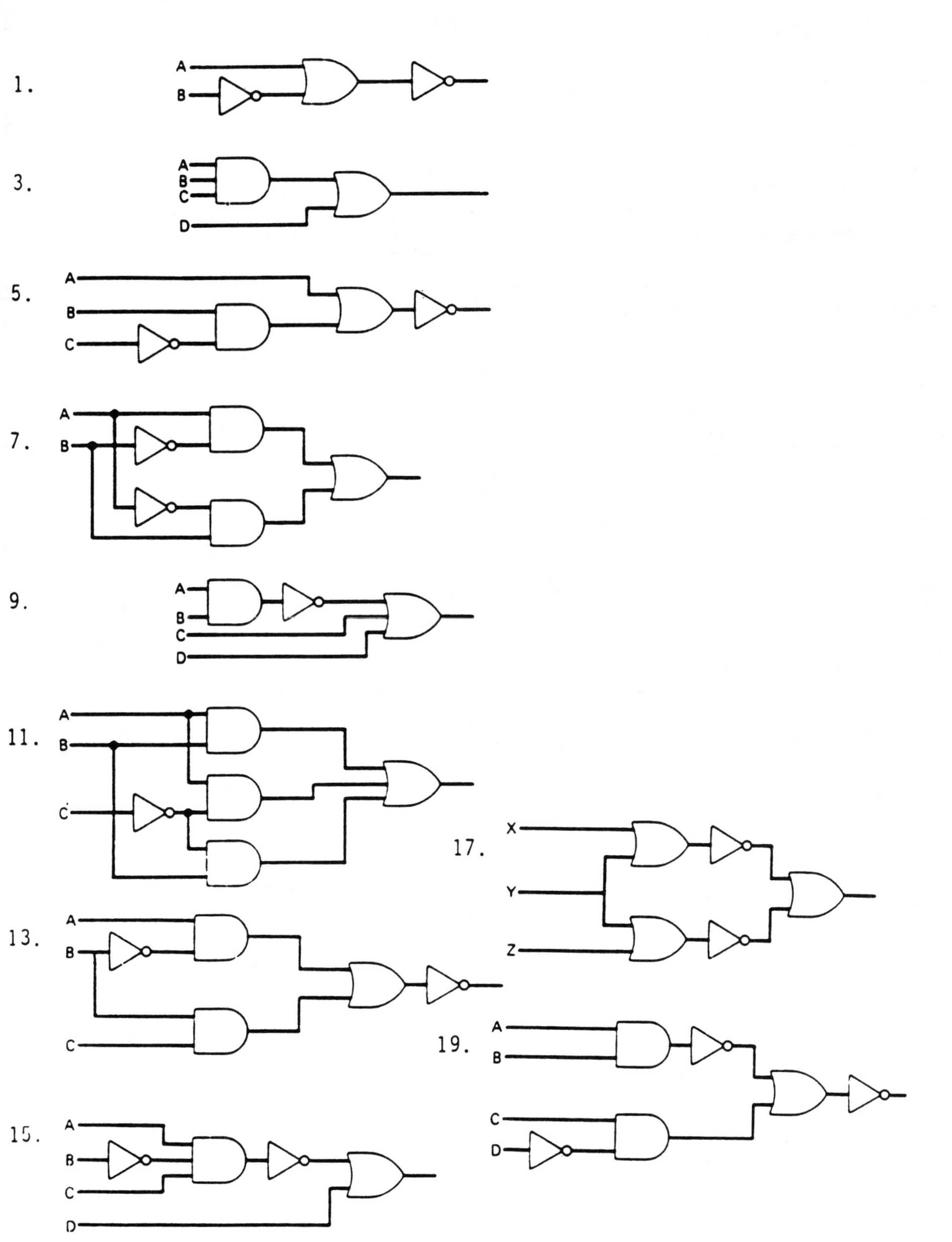

## *Problems 26-2*

1. $A\overline{B} + \overline{B}C$

3. $\overline{\overline{X}Y}Z$

5. $\overline{\overline{A}\,\overline{B}} + C$

7. $(A + B)(\overline{\overline{A} + B})D$

9. $Z(X + Y)$

11. $\overline{\overline{AB}(A + B)}$

13. $\overline{X + \overline{Y}} \cdot \overline{\overline{Y} + Z}$

15. $\overline{\overline{A + B}(B + C)(C + D)}$

17. $\overline{XY + \overline{YZ}}$

19. $\overline{A\overline{B} + \overline{B}C}$

## *Problems 26-3*

1.

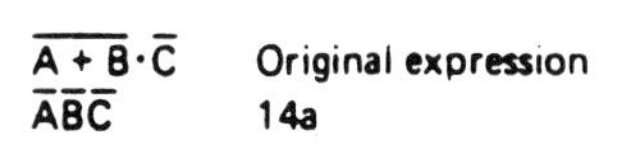


(a)

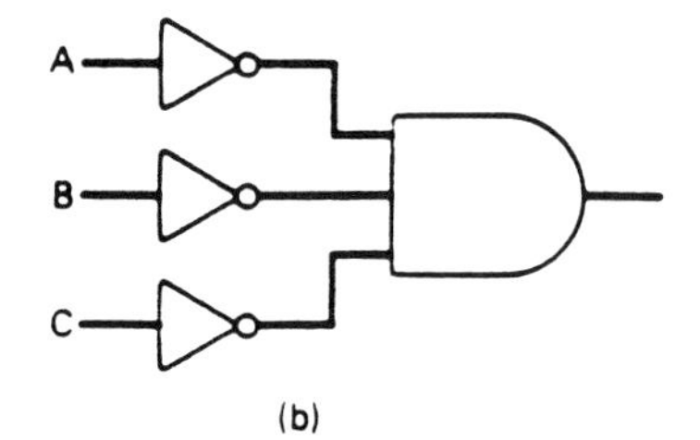


(b)

| A | B | C | $\overline{A}$ | $\overline{B}$ | $\overline{C}$ | A + B | $\overline{A + B}$ | $\overline{A + B} \cdot \overline{C}$ | $\overline{A}\,\overline{B}\,\overline{C}$ |
|---|---|---|---|---|---|---|---|---|---|
| 0 | 0 | 0 | 1 | 1 | 1 | 0 | 1 | 1 | 1 |
| 0 | 0 | 1 | 1 | 1 | 0 | 0 | 1 | 0 | 0 |
| 0 | 1 | 0 | 1 | 0 | 1 | 1 | 0 | 0 | 0 |
| 0 | 1 | 1 | 1 | 0 | 0 | 1 | 0 | 0 | 0 |
| 1 | 0 | 0 | 0 | 1 | 1 | 1 | 0 | 0 | 0 |
| 1 | 0 | 1 | 0 | 1 | 0 | 1 | 0 | 0 | 0 |
| 1 | 1 | 0 | 0 | 0 | 1 | 1 | 0 | 0 | 0 |
| 1 | 1 | 1 | 0 | 0 | 0 | 1 | 0 | 0 | 0 |

Equal

(c)

## Problems 26-3

**3.**

| | |
|---|---|
| $(A + B)(\overline{A + C})D$ | Original expression |
| $(A + B)(\bar{A}\bar{C})D$ | 14a |
| $A\bar{A}\bar{C}D + \bar{A}B\bar{C}D$ | 7a |
| $0 + \bar{A}B\bar{C}D$ | 12a |
| $\bar{A}B\bar{C}D$ | 9b |

(a)

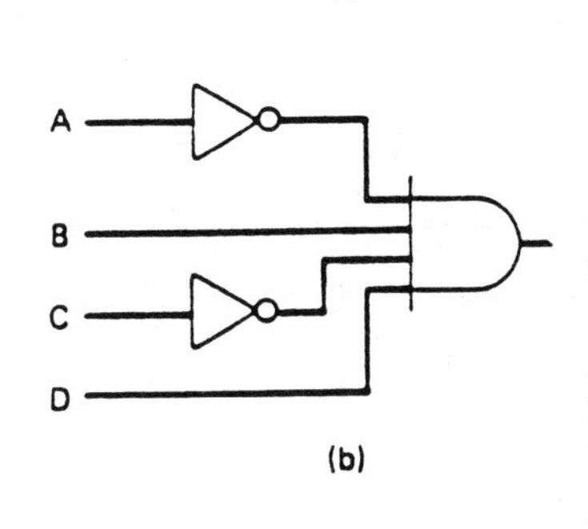


(b)

| A | B | C | D | $\bar{A}$ | $\bar{C}$ | A + B | A + C | $\overline{A + C}$ | $D(A + B)(\overline{A + C})$ | $\bar{A}B\bar{C}D$ |
|---|---|---|---|---|---|---|---|---|---|---|
| 0 | 0 | 0 | 0 | 1 | 1 | 0 | 0 | 1 | 0 | 0 |
| 0 | 0 | 0 | 1 | 1 | 1 | 0 | 0 | 1 | 0 | 0 |
| 0 | 0 | 1 | 0 | 1 | 0 | 0 | 1 | 0 | 0 | 0 |
| 0 | 0 | 1 | 1 | 1 | 0 | 0 | 1 | 0 | 0 | 0 |
| 0 | 1 | 0 | 0 | 1 | 1 | 1 | 0 | 1 | 0 | 0 |
| 0 | 1 | 0 | 1 | 1 | 1 | 1 | 0 | 1 | 1 | 1 |
| 0 | 1 | 1 | 0 | 1 | 0 | 1 | 1 | 0 | 0 | 0 |
| 0 | 1 | 1 | 1 | 1 | 0 | 1 | 1 | 0 | 0 | 0 |
| 1 | 0 | 0 | 0 | 0 | 1 | 1 | 1 | 0 | 0 | 0 |
| 1 | 0 | 0 | 1 | 0 | 1 | 1 | 1 | 0 | 0 | 0 |
| 1 | 0 | 1 | 0 | 0 | 0 | 1 | 1 | 0 | 0 | 0 |
| 1 | 0 | 1 | 1 | 0 | 0 | 1 | 1 | 0 | 0 | 0 |
| 1 | 1 | 0 | 0 | 0 | 1 | 1 | 1 | 0 | 0 | 0 |
| 1 | 1 | 0 | 1 | 0 | 1 | 1 | 1 | 0 | 0 | 0 |
| 1 | 1 | 1 | 0 | 0 | 0 | 1 | 1 | 0 | 0 | 0 |
| 1 | 1 | 1 | 1 | 0 | 0 | 1 | 1 | 0 | 0 | 0 |

(c)

**5.**

| | |
|---|---|
| $\overline{A\bar{B} + \bar{B}C}$ | Original expression |
| $\overline{A\bar{B}}\cdot\overline{\bar{B}C}$ | 14a |
| $(\bar{A} + B)(B + \bar{C})$ | 14b |
| $\bar{A}B + BB + \bar{A}\bar{C} + B\bar{C}$ | 8a |
| $\bar{A}B + B + \bar{A}\bar{C} + B\bar{C}$ | 11a |
| $B + \bar{A}\bar{C} + B\bar{C}$ | 15b |
| $B + B\bar{C} + \bar{A}\bar{C}$ | 6b |
| $B + \bar{A}\bar{C}$ | 15b |

(a)

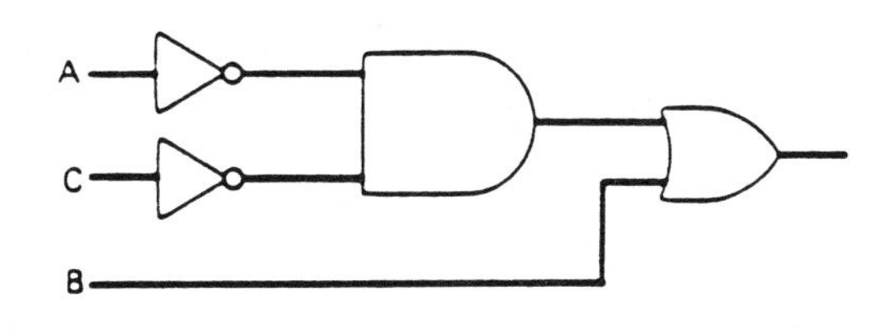


(b)

| A | B | C | $\bar{A}$ | $\bar{B}$ | $\bar{C}$ | $A\bar{B}$ | $\bar{B}C$ | $A\bar{B} + \bar{B}C$ | $\overline{A\bar{B} + \bar{B}C}$ | $\bar{A}\bar{C}$ | $B + \bar{A}\bar{C}$ |
|---|---|---|---|---|---|---|---|---|---|---|---|
| 0 | 0 | 0 | 1 | 1 | 1 | 0 | 0 | 0 | 1 | 1 | 1 |
| 0 | 0 | 1 | 1 | 1 | 0 | 0 | 1 | 1 | 0 | 0 | 0 |
| 0 | 1 | 0 | 1 | 0 | 1 | 0 | 0 | 0 | 1 | 1 | 1 |
| 0 | 1 | 1 | 1 | 0 | 0 | 0 | 0 | 0 | 1 | 0 | 1 |
| 1 | 0 | 0 | 0 | 1 | 1 | 1 | 0 | 1 | 0 | 0 | 0 |
| 1 | 0 | 1 | 0 | 1 | 0 | 1 | 1 | 1 | 0 | 0 | 0 |
| 1 | 1 | 0 | 0 | 0 | 1 | 0 | 0 | 0 | 1 | 0 | 1 |
| 1 | 1 | 1 | 0 | 0 | 0 | 0 | 0 | 0 | 1 | 0 | 1 |

Equal

(c)

## *Problems 26-3*

7.

| | |
|---|---|
| $\overline{A + B\overline{C}}$ | Original expression |
| $\overline{A}\,\overline{B\overline{C}}$ | 14a |
| $\overline{A}(\overline{B} + C)$ | 14b |

(a)

(b)

| A | B | C | $\overline{A}$ | $\overline{B}$ | $\overline{C}$ | $B\overline{C}$ | $A + B\overline{C}$ | $\overline{A + B\overline{C}}$ | $\overline{B} + C$ | $\overline{A}(\overline{B} + C)$ |
|---|---|---|---|---|---|---|---|---|---|---|
| 0 | 0 | 0 | 1 | 1 | 1 | 0 | 0 | 1 | 1 | 1 |
| 0 | 0 | 1 | 1 | 1 | 0 | 0 | 0 | 1 | 1 | 1 |
| 0 | 1 | 0 | 1 | 0 | 1 | 1 | 1 | 0 | 0 | 0 |
| 0 | 1 | 1 | 1 | 0 | 0 | 0 | 0 | 1 | 1 | 1 |
| 1 | 0 | 0 | 0 | 1 | 1 | 0 | 1 | 0 | 1 | 0 |
| 1 | 0 | 1 | 0 | 1 | 0 | 0 | 1 | 0 | 1 | 0 |
| 1 | 1 | 0 | 0 | 0 | 1 | 1 | 1 | 0 | 0 | 0 |
| 1 | 1 | 1 | 0 | 0 | 0 | 0 | 1 | 0 | 1 | 0 |

(c) Equal

9.

| | |
|---|---|
| $(\overline{B + C})(C + D)$ | Original expression |
| $\overline{B}\,\overline{C}(C + D)$ | 14a |
| $\overline{B}\,\overline{C}C + \overline{B}\,\overline{C}D$ | 8a |
| $\overline{B}\cdot 0 + \overline{B}\,\overline{C}D$ | 12a |
| $0 + \overline{B}\,\overline{C}D$ | 9a |
| $\overline{B}\,\overline{C}D$ | 9b |

(a)

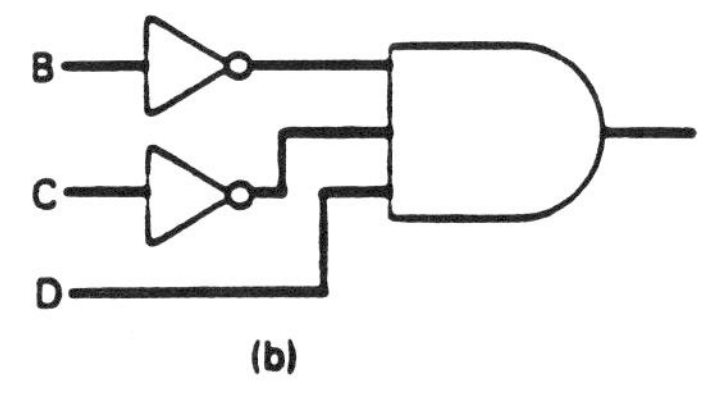


(b)

| B | C | D | $\overline{B}$ | $\overline{C}$ | B + C | $\overline{B + C}$ | C + D | $(\overline{B + C})(C + D)$ | $\overline{B}\,\overline{C}D$ |
|---|---|---|---|---|---|---|---|---|---|
| 0 | 0 | 0 | 1 | 1 | 0 | 1 | 0 | 0 | 0 |
| 0 | 0 | 1 | 1 | 1 | 0 | 1 | 1 | 1 | 1 |
| 0 | 1 | 0 | 1 | 0 | 1 | 0 | 1 | 0 | 0 |
| 0 | 1 | 1 | 1 | 0 | 1 | 0 | 1 | 0 | 0 |
| 1 | 0 | 0 | 0 | 1 | 1 | 0 | 0 | 0 | 0 |
| 1 | 0 | 1 | 0 | 1 | 1 | 0 | 1 | 0 | 0 |
| 1 | 1 | 0 | 0 | 0 | 1 | 0 | 1 | 0 | 0 |
| 1 | 1 | 1 | 0 | 0 | 1 | 0 | 1 | 0 | 0 |

(c) Equal

## *Problems 26-3*

**11.**

| Expression | Step |
|---|---|
| $\overline{\overline{AB} + C\bar{D}}$ | Original expression |
| $\overline{\overline{AB}}\cdot\overline{C\bar{D}}$ | 14a |
| $AB\cdot\overline{C\bar{D}}$ | 13a |
| $AB(\bar{C} + D)$ | 14b |

(a)

A B C D

(b)

| A | B | C | D | $\bar{C}$ | $\bar{D}$ | AB | $\overline{AB}$ | $C\bar{D}$ | $\overline{AB} + C\bar{D}$ | $\overline{\overline{AB} + C\bar{D}}$ | $\bar{C} + D$ | $AB(\bar{C} + D)$ |
|---|---|---|---|---|---|---|---|---|---|---|---|---|
| 0 | 0 | 0 | 0 | 1 | 1 | 0 | 1 | 0 | 1 | 0 | 1 | 0 |
| 0 | 0 | 0 | 1 | 1 | 0 | 0 | 1 | 0 | 1 | 0 | 1 | 0 |
| 0 | 0 | 1 | 0 | 0 | 1 | 0 | 1 | 1 | 1 | 0 | 0 | 0 |
| 0 | 0 | 1 | 1 | 0 | 0 | 0 | 1 | 0 | 1 | 0 | 1 | 0 |
| 0 | 1 | 0 | 0 | 1 | 1 | 0 | 1 | 0 | 1 | 0 | 1 | 0 |
| 0 | 1 | 0 | 1 | 1 | 0 | 0 | 1 | 0 | 1 | 0 | 1 | 0 |
| 0 | 1 | 1 | 0 | 0 | 1 | 0 | 1 | 1 | 1 | 0 | 0 | 0 |
| 0 | 1 | 1 | 1 | 0 | 0 | 0 | 1 | 0 | 1 | 0 | 1 | 0 |
| 1 | 0 | 0 | 0 | 1 | 1 | 0 | 1 | 0 | 1 | 0 | 1 | 0 |
| 1 | 0 | 0 | 1 | 1 | 0 | 0 | 1 | 0 | 1 | 0 | 1 | 0 |
| 1 | 0 | 1 | 0 | 0 | 1 | 0 | 1 | 0 | 1 | 0 | 0 | 0 |
| 1 | 0 | 1 | 1 | 0 | 0 | 0 | 1 | 0 | 1 | 0 | 1 | 0 |
| 1 | 1 | 0 | 0 | 1 | 1 | 1 | 0 | 0 | 0 | 1 | 1 | 1 |
| 1 | 1 | 0 | 1 | 1 | 0 | 1 | 0 | 0 | 0 | 1 | 1 | 1 |
| 1 | 1 | 1 | 0 | 0 | 1 | 1 | 0 | 1 | 1 | 0 | 0 | 0 |
| 1 | 1 | 1 | 1 | 0 | 0 | 1 | 0 | 0 | 0 | 1 | 1 | 1 |

Equal

(c)

# CHAPTER 27

***Problems 27-1***

1.

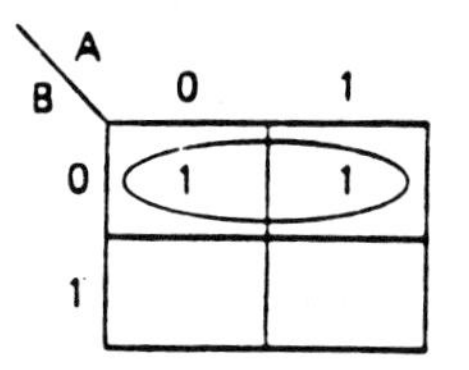


MAP = $\overline{B}$

3.

| C \ AB | 0, 0 | 0, 1 | 1, 1 | 1, 0 |
|---|---|---|---|---|
| 0 | 1 | 1 | | |
| 1 | | 1 | | |

MAP = $A\overline{C} + \overline{A}B$

5.

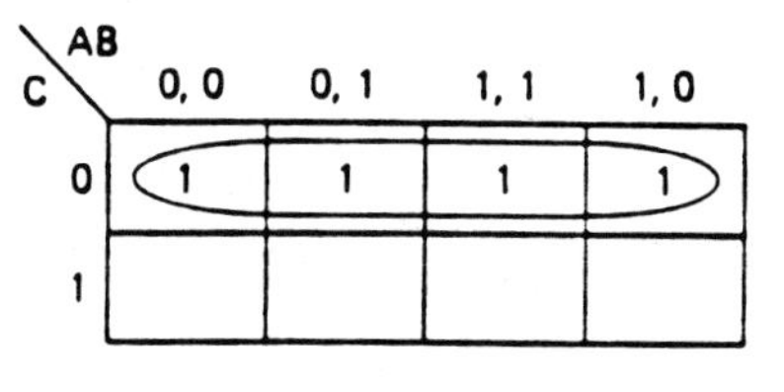


MAP = $\overline{C}$

7.

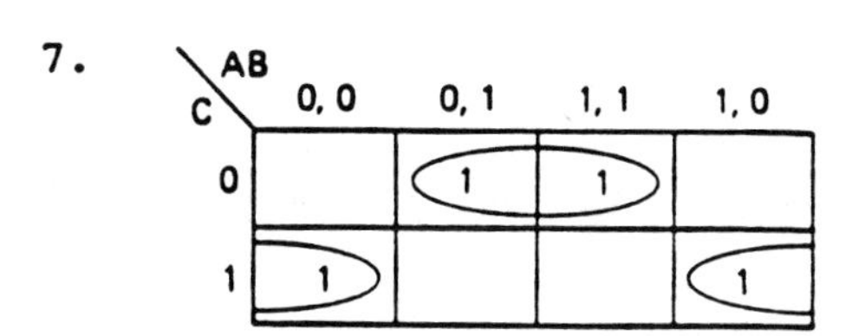


MAP = $B\overline{C} + \overline{B}C$

9. MAP = B

11. $\overline{C} + \overline{A}B$

***Problems 27-2***

1.

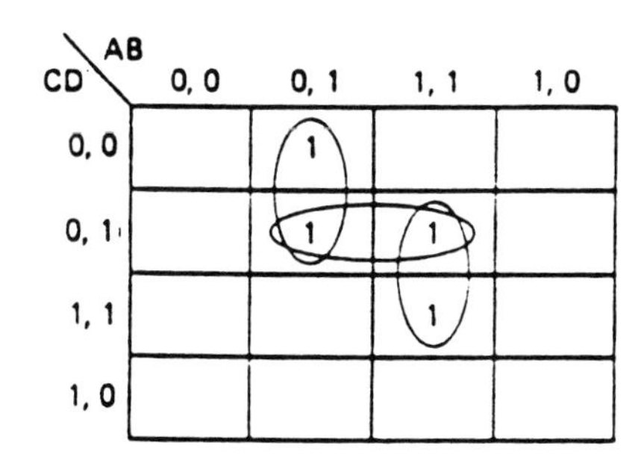


MAP = $\overline{A}B\overline{C} + B\overline{C}D + ABD$

3.

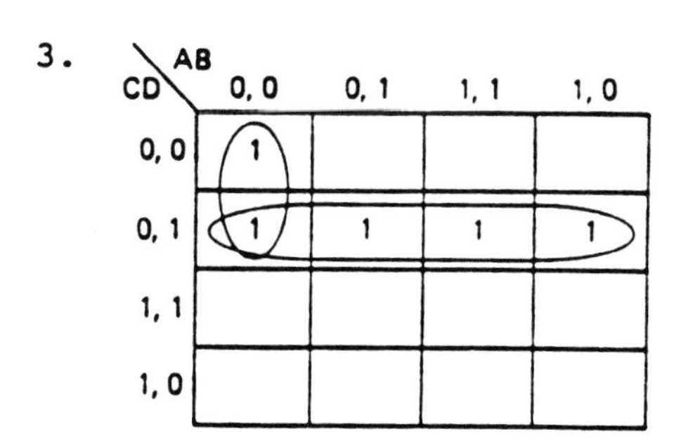


MAP = $\overline{C}D + \overline{A}\,\overline{B}\,\overline{C}$

5.

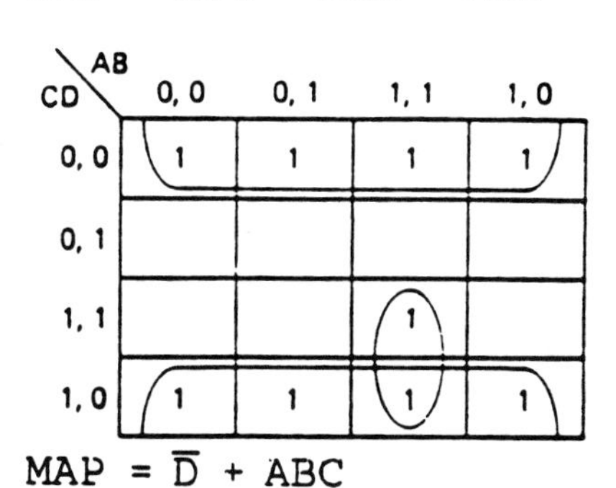


MAP = $\overline{D} + ABC$

7. MAP = D

9. MAP = $B\overline{C} + \overline{B}CD$

# CHAPTER 28

*Problems 28-1*

1. 15 scores on a quiz:

```
                          X
                          X
                          X
              X     X     X     X     X     X
        X     X     X     X     X     X     X
      60(1) 65(2) 70(2) 75(5) 80(2) 85(2) 90(1)
```

2. 15 scores on a quiz:

```
                          X
                          X
        X                 X
        X     X           X     X     X
        X     X           X     X     X     X
      60(3) 65(2)       70(5) 75(2) 80(2) 85(1)
```

```
      60(3) 65(2) 70(5) 75(2) 80(2) 85(1)
        X     X     X     X     X     X
        X     X     X     X     X
        X           X
                    X
                    X
```

3. 20 amplifiers supposed to be 40 dB each

```
                    X
                    X
                    X     X
  X           X     X     X
  X     X     X     X     X
  X     X     X     X     X     X     X
37(3) 38(2) 39(4) 40(6) 41(3) 42(1) 43(1)
```

intervals are divided by whole numbers: 37 = 37.00 to 37.99

4. 20 amplifiers supposed to be 40 dB each

```
                    X
                    X
                    X
  X     X     X     X     X
  X     X     X     X     X
  X     X     X     X     X     X     X
37(3) 38(3) 39(3) 40(6) 41(3) 42(1) 43(1)
```

intervals are divided by whole numbers: 37 = 37.00 to 37.99

5. 25 DAYS production

```
                                  X      X
        X          X      X       X      X
X       X     X    X      X       X      X      X      X
X       X     X    X      X       X      X      X      X
260(2) 270(3) 280(2) 290(3) 300(3) 310(4) 320(4) 330(2) 340(2)
```

intervals are divided by sets of ten: 260 = 260 to 269

6. 25 productive days over 30 DAYS

| | | | | | | |
|---|---|---|---|---|---|---|
| | X | | | | | |
| | X | X | | | | |
| | X | X | | | | |
| X | X | X | | | | |
| X | X | X | X | X | | |
| X | X | X | X | X | | |
| X | X | X | X | X | X | X |
| 80(4) | 90(7) | 100(6) | 110(3) | 120(3) | 130(1) | 140(1) |

intervals are divided by sets of ten: 80 = 80 TO 89

7. 25 samples in kΩ

| | | | | | | | | | |
|---|---|---|---|---|---|---|---|---|---|
| | | | | | X | | X | | |
| | | | | X | X | X | X | | |
| | | | | X | X | X | X | | |
| | | | X | X | X | X | X | X | X |
| X | | | X | X | X | X | X | X | X |
| 1.90 | 1.92 | 1.94 | 1.96 | 1.98 | 2.00 | 2.02 | 2.04 | 2.06 | 2.08 |
| (1) | (0) | (0) | (2) | (4) | (5) | (4) | (5) | (2) | (2) |

intervals are divided by sets of .02: 1.90 = 1.90 to 1.91

8. 25 samples in kΩ

| | | | | | | | | | | |
|---|---|---|---|---|---|---|---|---|---|---|
| | | | | X | X | | | | | |
| | | | | X | X | | | | | |
| | X | | | X | X | | | X | | |
| | X | | X | X | X | X | | X | | |
| X | X | | X | X | X | X | X | X | X | X |
| 44.5 | 45.0 | 45.5 | 46.0 | 46.5 | 47.0 | 47.5 | 48.0 | 48.5 | 49.0 | 49.5 |
| (1) | (3) | (0) | (2) | (5) | (5) | (2) | (1) | (3) | (1) | (1) |

intervals in sets of .5: 44.5 = 44.5 to 44.95

9. 20 days results of defective parts per hundred:

| | | | | | |
|---|---|---|---|---|---|
| | | X | | | |
| | | X | | | |
| | | X | | | |
| | | X | X | | |
| | X | X | X | X | |
| X | X | X | X | X | |
| X | X | X | X | X | X |
| 1(2) | 2(3) | 3(7) | 4(4) | 5(3) | 6(1) |

10. 30 days results
of defective parts
per hundred:

(day 28 = 4)

| | | | | | |
|---|---|---|---|---|---|
| | | | X | | |
| | | | X | | |
| | | | X | | |
| | | X | X | X | |
| | X | X | X | X | |
| | X | X | X | X | |
| | X | X | X | X | X |
| | X | X | X | X | X |
| X | X | X | X | X | X |
| 1(1) | 2(5) | 3(6) | 4(9) | 5(6) | 6(3) |

*Problems 28-2*

1. 15 scores on a quiz:
   60 65 65 70 70 75 75 75 75 75 80 80 85 85 90
   mean = 75, median = 75, mode = 75, $\sigma = 7.958$

2. 15 scores on a quiz:
   60 60 60 65 65 70 70 70 70 70 75 75 80 80 85
   mean = 70.33, median = 70, mode = 70, $\sigma = 7.41$

3. 20 amplifiers supposed to be 40 dB each:
   37.0 37.3 37.6 38.2 38.4 39.2 39.2 39.4 39.7 40.2
   40.3 40.4 40.7 40.7 40.9 41.3 41.6 41.8 42.6 43.6
   mean = 40.005, median = 40.25, mode = 39.2 and 40.7, $\sigma = 1.72$

4. 20 amplifiers supposed to be 40 dB each:
   37.4 37.6 37.6 38.1 38.4 38.8 39.4 39.7 39.8 40.2
   40.3 40.4 40.6 40.8 40.8 41.1 41.7 41.7 42.6 43.6
   mean = 40.03, median = 40.25, mode = 37.6 and 40.8 and 41.7,
   $\sigma = 1.66$

5. 25 DAYS number of items produced each day:
   260 265 270 276 277 288 288 290 293 295 300 302 306
   310 312 315 315 320 320 325 327 330 332 340 340
   mean = 303.84, median = 306,
   mode = 288 and 315 and 320 and 340, $\sigma = 22.75$

6. 25 productive days over 30 DAYS:
   80 85 88 89 93 94 94 95 96 98 98 102 103
   104 104 105 107 113 118 119 120 122 123 130 140
   mean = 104.8, median = 103, mode = 94 and 98 and 104, $\sigma = 14.66$

7. 25 samples in kΩ:
1.90 1.96 1.96 1.98 1.98 1.98 1.98 2.00 2.00 2.00
2.00 2.00 2.02 2.02 2.03 2.03 2.04 2.04 2.04 2.05
2.05 2.06 2.06 2.08 2.09
mean = 2.014, median = 2.02, mode = 2.00 , $\sigma$ = .04195

8. 25 sample in kΩ:
44.7 45.2 45.2 45.3 46.0 46.2 46.5 46.5 46.7 46.8
46.8 46.9 47.0 47.0 47.1 47.2 47.3 47.5 47.8 48.1
48.5 48.6 48.7 49.0 49.8
mean = 47.056, median = 47.0,
mode = 45.2 and 46.5 and 46.8 and 47.0, $\sigma$ = 1.24, Sx = 1.27

9. 20 days results of defective parts per hundred:
1 1 2 2 2 3 3 3 3 3 3 3 4 4 4 4 5 5 5 6
mean = 3.3, median = 3, mode = 3, $\sigma$ = 1.308, Sx =1.342

10. 30 days results of defective parts per hundred: (day 28 = 4)
1 2 2 2 2 2 3 3 3 3 3 3 4 4 4
4 4 4 4 4 4 5 5 5 5 5 5 6 6 6
mean = 3.77, median = 4, mode = 4, $\sigma$ = 1.31

*Problems 28-3*

1. 60(1) 65(3) 70(5) 75(7) 80(5) 85(4) 90(2) 95(1) 100(1)
n = 9, mean = 77.4, median = 75, mode = 75,
range = 100-60 = 40, standard deviation = 9.25

2. 60(1) 65(2) 70(3) 75(5) 80(10) 85(6) 90(4) 95(2) 100(1)
n = 34, mean = 80.44, median = 80, mode = 80,
range = 100-60 = 40, standard deviation = 8.86

3. 55(2) 60(1) 65(4) 70(3) 75(6) 80(10) 85(5) 90(4) 95(2)
n = 37, mean = 77.43, median = 80, mode = 80,
range = 95-55 = 40, standard deviation = 10.04

4. 55(1) 60(0) 65(3) 70(6) 75(7) 80(9) 85(5) 90(2) 95(1)
n = 34, mean = 76.92, median = (75+80)/2 = 77.5, mode = 80,
range = 95-55 = 40, standard deviation = 8.23

5. 50-56(2) 57-63(5) 64-70(7) 71-77(8) 78-84(6) 85-91(4) 92-98(2)
Use midpoint in any interval as the data value for that interval.
Use 54(2) 60(5) 67(7) 74(8) 81(6) 88(4) 95(2)
n = 34, mean = 73.44, median = 74, mode = 74,
range = 98-50 = 48, standard deviation = 10.94

6. 50-56(1) 57-63(2) 64-70(5) 71-77(8) 78-84(10) 85-91(7) 92-98(4)
use 54(1) 60(2) 67(5) 74(8) 81(10) 88(7) 95(4)
n = 37, mean = 78.57, median = 81, mode = 81,
range = 98-50 = 48, standard deviation = 10.25

7. 50-54(2) 55-59(5) 60-64(5) 65-69(3) 70-74(4) 75-79(6)
80-84(8) 85-89(3) 90-94(1)
Use 52(2) 57(5) 62(5) 67(3) 72(4) 77(6) 82(8) 87(3) 92(1)
n = 37, mean = 71.86, median = 72, mode = 82,
range = 94-50 = 44, standard deviation = 11.06

8. 50-54(1) 55-59(4) 60-64(7) 65-69(4) 70-74(3) 75-79(4)
80-84(10) 85-89(5) 90-94(2)
Use 52(1) 57(4) 62(7) 67(4) 72(3) 77(4) 82(10) 87(5) 92(2)
n = 40, mean = 73.625, median = 77, mode = 82,
range =, standard deviation = 11.20

9. 300-340(1) 341-381(3) 382-422(5) 423-463(8)
464-504(12) 505-545(6) 546-586(4) 587-627(2)
Use 320(1) 361(3) 402(5) 443(8) 484(12) 525(6) 566(4) 607(2)
n = 41, mean = 473, median = 484, mode = 484,
range = 627-300 = 327, standard deviation = 66.56

10. 710-760(2) 761-811(3) 812-862(7) 863-913(11)
914-964(7) 965-1015(5) 1016-1066(3) 1067-1117(1)
Use 735(2) 786(3) 837(7) 888(11) 939(7) 990(5) 1041(3) 1092(1)
n = 39, mean = 902.38, median = 888, mode = 888,
range = 1117-710 = 407, standard deviation = 83.24

11. 500-520(4) 521-541(6) 542-562(11) 563-583(7)
584-604(5) 605-625(3) 626-646(1)
Use 510(4) 531(6) 552(11) 573(7) 594(5) 615(3) 636(1)
n = 37, mean = 561.08, median = 552, mode = 552,
range =, standard deviation = 31.84

12. 140-170(2) 171-201(3) 202-232(4) 233-263(6) 264-294(10)
295-325(7) 326-356(4) 357-387(2) 388-418(1)
Use 155(2) 186(3) 217(4) 248(6) 279(10) 310(7) 341(4) 372(2) 413(1)
n = 39, mean = 274.49, median = 279, mode = 279,
range = 418-140 = 278, standard deviation = 58.70

*Problems 28-4*

1. one $\sigma$ range is determined by mean 40 mm $\pm$ 1 mm.
34 widgets (68% of 50) should be in the range of 39mm to 41mm.
47 widgets (95% of 50) should be in the range of 38mm to 42mm.
50 widgets (99.7% of 50) should be in the range of 37mm to 43mm.

2. one $\sigma$ range is determined by mean 200 mm $\pm$ 1.5 mm.
136 widgets (68% of 200) should be 198.5mm to 201.5mm.
190 widgets (95% of 200) should be 197mm to 203mm.
199 widgets (99.7% of 200) should be 194mm to 206mm.

3. one $\sigma$ range is determined by mean 600 parts $\pm$ 10 parts
Assuming round the clock production for 30 days of June,
68% of 30 = 20 days should have production of 590 to 610 parts.
95% of 30 = 29 days should have production of 580 to 620 parts.
99.7% of 30 = 30 days should have production of 570 to 630 parts.

4. one $\sigma$ range is determined by mean 400 parts $\pm$ 7.5 parts
Assuming round the clock production for 30 days of June,
68% of 30 = 20 days should produce 392.5 to 407.5 parts.
95% of 30 = 29 days should produce 385 to 415 parts.
99.7% of 30 = 30 days should produce 377.5 to 422.5 parts.

5. one $\sigma$ range is determined by mean score 73 $\pm$ 5.6
68% of 5000 = 3400 scores should be between 67.4 and 134.8
95% of 5000 = 4750 scores should be between 61.8 and 196.6
99.7% of 5000 = 4985 scores should be between 56.2 and 258.4

6. one $\sigma$ range is determined by mean score 77.5 $\pm$ 6.3
68% of 4000 = 2720 scores should be between 71.2 and 83.8
95% of 4000 = 3800 scores should be between 64.9 and 90.1
99.7% of 4000 = 3988 scores should be between 58.6 and 96.4